THE
FOUNDRYMAN'S
HANDBOOK

THE FOSECO
FOUNDRYMAN'S
HANDBOOK

FACTS

FIGURES AND

FORMULAE

* Reg. Trade Mark

Revised by

FOUNDRY SERVICES LIMITED,

Tamworth, Staffordshire, B78 3TL
Telephone: Tamworth 4141

PERGAMON PRESS
OXFORD · NEW YORK · TORONTO
SYDNEY · PARIS · FRANKFURT

U.K.	Pergamon Press Ltd., Headington Hill Hall, Oxford OX3 0BW, England.
U.S.A.	Pergamon Press Inc., Maxwell House, Fairview Park, Elmsford, New York 10523, U.S.A.
CANADA	Pergamon of Canada Ltd., 75 The East Mall, Toronto, Ontario, Canada
AUSTRALIA	Pergamon Press (Aust.) Pty. Ltd., 19a Boundary Street, Rushcutters Bay, N.S.W. 2011, Australia
FRANCE	Pergamon Press SARL, 24 rue des Ecoles, 75240 Paris, Cedex 05, France.
FEDERAL REPUBLIC OF GERMANY	Pergamon Press GmbH, 6242 Kronberg-Taunus, Pferdstrasse 1, Federal Republic of Germany.

Eighth Revised and Enlarged Edition 1975
Reprinted 1978

Library of Congress Cataloging in Publication Data

Foseco Services Limited.
The Foseco foundryman's handbook.

Sixth-7th ed. by Information Dept. of Foseco
International Limited.
1. Founding-Handbooks, manuals, etc. I. Foseco
International Limited. Information Dept. The Foseco
foundryman's handbook. II. Title.
TS235.F59 1975   671.2'02'02   74-19212
ISBN 0-08-018020-5

Printed in Great Britain by A. Wheaton & Co. Ltd., Exeter.

FOREWORD

This reference book has been prepared to help all those concerned in the making of castings. It is hoped that the information given will prove accurate and useful.

On any special problem not dealt with in this Handbook, the Technical Department of Foundry Services Limited is always available for consultation. Authoritative opinion and advice can be obtained from specialists in practically every aspect of foundrywork and no charge is normally made for this service.

The facts, figures, and formulae appearing in the following pages have been put together with great care, but should the reader discover any error or have constructive criticism or comment to offer we should be pleased to hear from you. A major attempt has been made to utilise metric measurements and SI units whenever possible. However, not all specifications, etc., have been re-issued in metric terms and some information, perforce, remains in avoirdupois and other units where direct conversion is not possible. Suggestions for material to be included in future editions would be welcomed and will be given very careful consideration.

PRINCIPAL CONTENTS

See end of Handbook for general subject index

SECTION I

TABLES AND GENERAL DATA

The tables appearing on pages 24-41, 44-53 and 60-61 are reproduced by kind permission of ROLLS ROYCE LIMITED, from their publication T.S.D.201 ©

THE INTERNATIONAL SYSTEM OF UNITS (SI) AND THEIR RELATIONSHIP TO UNITED KINGDOM AND OTHER MEASURES

The International System of Units, officially abbreviated SI, is a modernised version of the metric system established by international agreement. It is built upon a foundation of six base-units and their definitions which appear on the following pages. All other SI units are derived from these base-units. Multiples and sub-multiples are expressed in a decimal system.

Common equivalents and conversions to the more usual imperial weights and measures are given in the individual base-unit and also in a conversion table.

Acknowledgements: Ministry of Technology.
National Physical Laboratory.
United States Department of Commerce.
National Bureau of Standards.

LENGTH   METRE—m

The metre is defined as 1 650 763.73 wavelengths in vacuum of a specified orange-red line of the spectrum of krypton-86 atom.

The SI unit of area is the *square metre* (m²). Land is often measured by the hectare (10 000 square metres, or approximately 2.5 acres).

Metre bar

1 650 763·73 wavelengths

One wavelength

An interferometer is used to measure length by means of light waves

Kr⁸⁶atom

The SI unit of volume is the *cubic metre* (m³). Fluid volume is often measured by the litre (0.001 cubic metre).

TIME   SECOND—s

The second is defined as the duration of 9 192 631 770 cycles of radiation associated with a specified transition of the caesium-133 atom. It is realised by tuning an oscillator to the resonance frequency of the caesium atoms as they pass through a system of magnets and a resonant cavity into a detector.

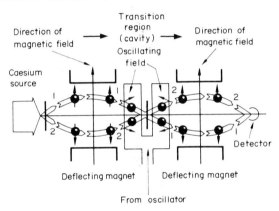

A schematic of an atomic beam clock. The trajectories are drawn for those atoms whose magnetic moments are reversed in the transition region.

The number of periods or cycles per second is called frequency. The SI unit for frequency is the *hertz* (Hz). One hertz equals one cycle per second.

Standard frequencies and correct time are broadcast from GPO station Rugby on frequencies 16, 60 kilohertz, 2.5, 5, 10 megahertz, and the BBC Droitwich transmitter 200 kilohertz. The standard radio broadcast band extends from 525 to 1605 kilohertz (medium frequency broadcasting band).

Dividing distance by time gives speed. The SI unit for speed is the *metre per second* (m/s), approximately 3 feet per second.

Rate of change in speed is called acceleration. The SI unit for acceleration is the *metre per second* squared (m/s²).

9

MASS   KILOGRAM–kg

The standard for the unit of mass, the kilogram, is a cylinder of platinum—iridium alloy kept by the International Bureau of Weights and Measures at Paris. A duplicate in the custody of the National Physical Laboratory — Teddington, serves as the mass standard for the United Kingdom. This is the only base-unit still defined by a material standard.

Kilogram

Closely allied to the concept of mass is that of force. The SI unit of force is the newton (N). A force of 1 newton, when applied for 1 second, will give to a 1 kilogram mass a speed of 1 metre per second (an acceleration of 1 metre per second squared).

1 Newton

Acceleration of 1m/s^2

$$1 N = \frac{1 kg \times 1m}{1 s^2}$$

One newton equals approximately two-tenths of a pound-force.
The weight of an object is the force exerted on it by gravity. Gravity gives a mass a downward acceleration of about 9.8 m/s^2.
The SI unit for work and energy of any kind is the *joule* (J).

$$1J = 1N \times 1m$$

The SI unit for power of any kind is the *watt* (W).

$$1W = \frac{1J}{1s}$$

TEMPERATURE   KELVIN—K

The thermodynamic or Kelvin scale of temperature used in SI has its origin or zero point at absolute zero and has a fixed point at the triple point of water defined as 273.16 kelvin. The Celsius* scale is derived from the Kelvin scale. The triple point is defined as 0.01°C on the Celsius scale, which is approximately 32.02°F on the Fahrenheit scale. The relationship of the Kelvin, Celsius, and Fahrenheit temperature scales is shown below.

Temp. F + 40 = 1·8(Temp. C + 40)
Temp. F = 1·8(Temp. C) + 32
Temp. C = (Temp. F − 32)/1·8
Temp. K = Temp. C + 273·15

The triple-point cell, an evacuated glass cylinder filled with pure water, is used to define a known fixed temperature. When the cell is cooled and some ice is formed in the water, the temperature at the interface of solid, liquid, and vapour is 0·01°C. Thermometers to be calibrated are placed in the re-entrant well.

*The metric practical unit of temperature formerly Centigrade.

11

LUMINOUS INTENSITY   CANDELA—cd

The candela is defined as the luminous intensity of 1/600 000 of a square metre of a radiating cavity at the temperature of freezing platinum under a specified pressure (about 2045 K).

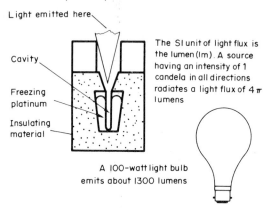

Light emitted here

Cavity

Freezing platinum

Insulating material

The SI unit of light flux is the lumen (lm). A source having an intensity of 1 candela in all directions radiates a light flux of 4π lumens

A 100-watt light bulb emits about 1300 lumens

These Prefixes may be Applied to all SI Units

Multiples and Submultiples	Prefixes	Symbols
$1\ 000\ 000\ 000\ 000 = 10^{12}$	tera (ter'*a*)	T
$1\ 000\ 000\ 000 = 10^{9}$	giga (ji'g*a*)	G
$1\ 000\ 000 = 10^{6}$	mega (meg'*a*)	M*
$1\ 000 = 10^{3}$	kilo (kil'o)	k*
$100 = 10^{2}$	hecto (hek'to)	h
$10 = 10$	deca (dek'*a*)	da
$0 \cdot 1 = 10^{-1}$	deci (des'i)	d
$0 \cdot 01 = 10^{-2}$	centi (sen'ti)	c
$0 \cdot 001 = 10^{-3}$	milli (mil'i)	m*
$0 \cdot 000\ 001 = 10^{-6}$	micro (mi'kro)	μ*
$0 \cdot 000\ 000\ 001 = 10^{-9}$	nano (nan'o)	n
$0 \cdot 000\ 000\ 000\ 001 = 10^{-12}$	pico (pe'ko)	p
$0 \cdot 000\ 000\ 000\ 000\ 001 = 10^{-15}$	femto (fem'to)	f
$0 \cdot 000\ 000\ 000\ 000\ 000\ 001 = 10^{-18}$	atto (at'to)	a

*Most commonly used

ELECTRIC CURRENT   AMPERE—A

The ampere is defined as the constant current that, when flowing through each of two long, thin parallel wires separated by one metre in free space, results in a force between the two wires (due to their magnetic fields) of 2×10^{-7} newton for each metre of length. (Simplified version.)

The SI unit of voltage is the *volt* (V).

$$1V = \frac{1W}{1A}$$

The SI unit of electrical resistance is the *ohm* (Ω).

$$1\Omega = \frac{1V}{1A}$$

Common Equivalents and Conversions

Approximate common equivalent		Conversions accurate to parts per million	
1 inch	= 25 millimetres	inches × 25·4	= millimetres
1 foot	= 0·3 metre	feet × 0·304 8	= metres
1 yard	= 0·9 metre	yards × 0·914 4	= metres
1 mile	= 1·6 kilometres	miles × 1·609 34	= kilometres
1 square inch	= 650 square millimetres	square inches × 645·16	= square millimetres
1 square foot	= 0·09 square metre	square feet × 0·092 903 0	= square metres
1 square yard	= 0·84 metre	square yards × 0·836 127	= square metres
1 acre	= 4047 square metres	acres × 4046·86	= square metres
1 cubic inch	= 16 cubic centimetres	cubic inches × 16·387 1	= cubic centimetres
1 cubic foot	= 0·028 cubic metre	cubic feet × 0·028 316 8	= cubic metres
1 cubic yard	= 0·8 cubic metre	cubic yards × 0·764 555	= cubic metres
1 pint	= 0·568 litre	pints × 0·568 261	= litres or cubic decimetres
1 gallon	= 4·5 litres	gallons × 4·546 09	= litres
1 ounce	= 28 grams	ounces (avdp) × 28·3495	= grams
1 pound	= 0·45 kilograms	pounds (avdp) × 0·453 592	= kilograms
1 cwt	= 51 kilograms	cwt × 50·802 3	= kilograms
1 ton	= 1016 kilograms or 1·016 tonnes	tons × 1016·05	= kilograms
		tons × 1·016 05	= tonnes (or metric tons)
1 horsepower	= 0·75 kilowatt	horsepower × 0·745 700	= kilowatts

14

1 millimetre	= 0·04 inch	
1 metre	= 3·3 feet	
1 metre	= 1·1 yards	
1 kilometre	= 0·6 mile	
1 square millimetre	= 0·0015 square inch	
1 square metre	= 11 square feet	
1 square metre	= 1·2 square yards	
1 square kilometre	= 247 acres	
1 cubic metre	= 35 cubic feet	
1 litre	= 0·22 gallon	
1 gram	= 0·035 ounce (avdp)	
1 kilogram	= 2·2 pounds (avdp)	

millimetres × 0·039 370 1	= inches	
metres × 3·280 84	= feet	
metres × 1·093 61	= yards	
kilometres × 0·621 371	= miles	
square millimetres × 0·001 550	= square inches	
square metres × 10·763 9	= square feet	
square metres × 1·195 99	= square yards	
square kilometres × 247·105	= acres	
cubic metres × 35·314 7	= cubic feet	
litres × 0·219 69	= gallons	
grams × 0·035 274	= ounces	
kilograms × 2·204 62	= pounds	

TABLE OF PHYSICAL PROPERTIES OF METALS

Element	Symbol	Atomic Weight	Melting Point °C.	Boiling Point °C.	Latent Heat of Fusion cal./gm.	Specific Heat cal./gm./°C.
Aluminium	Al	26·97	660	1800	92·4	0·2096
Antimony	Sb	121·76	630	1645	24·3	0·0495
Arsenic	As	74·93	Volatilises	S'blimes450	—	0·0758
Barium	Ba	137·37	704	1000 appr.	—	0·068
Beryllium	Be	9·02	1281	1500	31·89	0·425
Bismuth	Bi	209·0	269	1560	13·0	0·0304
Cadmium	Cd	112·41	321	767	14·0	0·0547
Calcium	Ca	40·08	851	1170	78·5	0·145
Carbon	C	12·005	3500	—	—	0·168
Cerium	Ce	140·13	804	1400	—	0·0447
Chromium	Cr	52·01	1830	2260	31·70	0·104
Cobalt	Co	58·94	1490	3467	58·4	0·103
Copper	Cu	63·57	1083	2325	43·0	0·0909
Gallium	Ga	69·74	29·8	2000	19·16	0·079
Gold	Au	197·2	1063	2530	16·1	0·0316
Indium	In	114·8	155	1450	—	0·0570
Iridium	Ir	193·1	2350	4800	—	0·0323
Iron	Fe	55·84	1530	3235	47·9	0·1045
Lead	Pb	207·22	327	1755	5·0	0·0302
Lithium	Li	6·94	186	1400	32·81	0·837
Magnesium	Mg	24·32	659	1107	46·5	0·246
Manganese	Mn	54·93	1242	1900	36·5	0·122
Mercury	Hg	200·61	−38·80	356·7	3·0	0·0333

Table of Physical Properties of Metals—continued

Element	Symbol	Atomic Weight	Melting Point °C.	Boiling Point °C.	Latent Heat of Fusion cal./gm.	Specific Heat cal./gm./°C.
Molybdenum	Mo	96.0	2622	3550	—	0.0659
Nickel	Ni	58.69	1454.9	3075	73.0	0.103
Niobium	Nb	92.91	2500	3700	—	—
Osmium	Os	190.9	2700	—	—	0.0311
Palladium	Pd	106.7	1549	2800	36.0	0.0592
Phosphorus (Yellow) ...	P	31.04	44.1	279	5.0	0.189
Platinum	Pt	195.23	1774	4300	27.0	0.0324
Potassium	K	39.10	63.6	758	16.0	0.187
Rhodium	Rh	102.91	1966	2520	—	0.058
Silicon	Si	28.3	1415	2392	120	0.123
Silver	Ag	107.88	960.5	2150	22	0.0556
Sodium	Na	22.997	97.6	877	27.5	0.2829
Strontium... ...	Sr	87.63	771	1366	—	0.0550
Sulphur (rhombic) ...	S	32.0	115	444.5	9.0	0.0163
Tantalum	Ta	180.89	2996	1390	37.0	0.0301
Tellurium	Te	127.6	450	1457	7.4	0.0525
Thallium	Tl	204	302	2270	—	0.0326
Tin	Sn	118.7	232	3000	14.6	0.0536
Titanium	Ti	47.9	1800	4830	90.0	0.122
Tungsten	W	184.0	3400	—	40.0	0.034
Uranium	U	238.2	1300	—	—	0.028
Vanadium	V	50.95	1720	913	80.0	0.1153
Zinc	Zn	65.38	419.4	2900	26.3	0.0918
Zirconium	Zr	90.6	1857	—	—	0.0660

Table of Physical Properties of Metals—continued

Element	Thermal Conductivity gm.-cals./cm. sec./°C.	Electrical Resistivity microhms cm.³	Change in Volume on Melting %	Density gm./cm.³	Thermal Co-efficient Linear Expansion x 10⁻⁶	Brinell Hardness No.
Aluminium …	0·480	2·65	6·6	2·70	23·86	17
Antimony …	0·044	39·00	1·4	6·68	11·52	30·0
Arsenic …	—	35·00	—	5·73	5·59	—
Barium …	—	—	—	3·75	—	—
Beryllium …	0·393	18·50	—	1·93	12·3	—
Bismuth …	0·0194	115·00	—3·32	9·80	13·46	9·0
Cadmium …	0·222	7·50	4·74	8·64	30·69	20
Calcium …	—	10·50	—	1·54	25·0	13
Carbon …	0·039	—	—	2·30	7·9	—
Cerium …	—	—	—	6·92	—	—
Chromium …	0·65	13·1	—	7·10	8·1	350
Cobalt …	0·1654	9·71	—	8·60	12·36	125
Copper …	0·941	1·69	4·05	8·94	16·6	48
Gallium …	—	57·1	—	5·95	18·3	—
Gold …	0·700	2·42	5·2	19·32	14·43	18·5
Indium …	0·057	9·00	—	7·31	33·0	1·0
Iridium …	0·141	5·30	—	22·41	7·00	172
Iron …	0·176	10·00	5·5	7·86	11·82	66
Lead …	0·083	20·80	3·4	11·37	29·24	—
Lithium …	0·1673	8·40	1·5	0·534	60·0	5·5
Magnesium …	0·376	4·35	4·2	1·74	26·94	25
Manganese …	—	5·00	—	7·39	22·8	—
Mercury …	0·0148	21·30	3·75	13·56	182·0	—

Table of Physical Properties of Metals—continued

Element	Thermal Conductivity gm.-cals./cm. sec./°C.	Electrical Resistivity microhms cm.³	Change in Volume on Melting %	Density gm./cm.³	Thermal Co-efficient Linear Expansion × 10⁻⁶	Brinell Hardness No.
Molybdenum	0·3489	4·77	—	10·00	5·01	147
Nickel	0·1428	6·40	—	8·90	12·79	80
Osmium	—	11·0	—	22·48	6·8	50
Palladium	0·17	10·70	—	11·4	11·76	0·6
Phosphorus (Yellow)	—	10·50	—	1·83	6·24	52
Platinum	0·166	6·64	2·8	21·50	8·99	0·04
Potassium	0·2365	5·10	—	0·862	83·0	156
Rhodium	0·2101	—	—	12·44	8·5	—
Silicon	0·20	—	4·5	2·30	7·63	25
Silver	1·006	1·62	2·5	10·50	19·3	0·10
Sodium	0·3225	4·74	—	0·971	72·0	—
Strontium	—	25·0	—	2·54	70·0	—
Sulphur (rhombic)	0·65	—	—	2·07	8·0	40
Tantalum	0·130	14·6	—	16·60	16·80	—
Tellurium	0·0144	21·0	—	6·25	30·21	—
Thallium	—	—	2·8	11·88	22·34	—
Tin	0·155	11·30	—	7·29	7·14	—
Titanium	—	3·0	—	4·50	4·44	—
Tungsten	0·35	5·48	—	19·30	—	—
Uranium	—	—	—	18·7	—	—
Vanadium	—	26·0	—	6·0	—	—
Zinc	0·265	6·1	6·5	7·10	30	35
Zirconium	—	—	—	6·5	—	—

CORRODIBILITY OF SOME COMMON METALS AND ALLOYS

Estimates are relative only: service temperatures are usually the deciding factors in designing equipment.

Originally compiled (1933) by Jerome Strauss; revised (1946) by E. E. Thum using data furnished by several specialists.

Class of Material	Nominal Composition						Atmospheres		Domestic	Water				Wet Steam
	Carbon	Chromium	Nickel	Silicon	Copper	Other Constituents	Sea Shore	Industrial		Mine	Sea	Saline with H₂S	Brackish with MD	
Ingot iron or wrought iron	0·03				0·08		P	P	F	P	F	F	F	FG
Low C steel	0·10				0·08		P	P	F	P	F	F	F	FG
Copper and high tensile steels	0·10				0·25		P	P	F	P	F	F	F	FG
Hot galvanized iron and steel							G	G	G	F	F	F	FG	FG
Calonized iron and steel							E	E	G	F	G	G	G	F
Grey cast iron	2-8 graphite, 0-7 combined carbon						F	F	G	F	F	F	F	F
High silicon iron	0·60			14·25			E	E	E	E	E	E	E	E
Nickel cast iron	3·30		3·50	2·00			F	F	G	F	FG	FG	FG	G
Chromium cast iron	2·50	25·00		2·00			G	G	G	G	G	G	G	G
Ni-Cr-Cu cast iron	3·00	2·00	14·00	1·50	6·00		FG	FG	G	FG	FG	FG	FG	G
Nickel steel: Low Ni	0·18		3·00				P	P	F	P	F	F	F	FG
High Ni	0·30		28·00				F	F	G	F	G	G	G	G
Chromium steels: 5% Cr.	0·15	5·00				Mo 0·5	F	F	G	F	F	F	F	G
7% Cr.	0·15 max.	7·00				Mo 0·5	G	G	G	G	G	G	G	G
9% Cr.	0·15 max.	9·00				Mo 1·0	G	G	G	G	G	G	G	G
12% Cr.	0·10 max.	12·00		0·75			G	G	E	G	G	G	G	G
17% Cr.	0·10 max.	17·00		0·50			GE	G	E	G	PF	PF	PF	E
17% Cr., 4% Mo	0·10 max.	17·00		0·75		Mo 4·0, Ni 0·25	E	E	E	E	FG	FG	FG	E
Silcrome (8% Cr, 3% Si)	0·35 max.	8·25		3·50			G	G	E	F	F	F	F	E
Cr-Ni steels: 8-20	0·45	8·00	20·00				E	E	E	E	PC	PC	PC	E
18-12, 2½% Mo	0·20	18·00	12·00			Mo 2·5	E	E	E	E	FG	FG	PF	GE
18-8	0·15	18·00	8·00				E	E	E	E	PC	PC	PC	E
18-12	0·10	18·00	12·00				E	E	E	E	FE	FE	FE	E
18-35	0·50 max.	18·00	35·00	1·00			E	E	E	E	PG	PG	PG	E
25-12	0·25 max.	25·00	12·00				E	E	E	E	PC	PC	PC	E
25-20	0·25 max.	25·00	20·00	1·50			E	E	E	E	G	G	G	E

E—Excellent; almost unlimited service.
G—Good; will give good service.
F—Fair.
P—Poor.

§Below 50° C.   ║May crack under certain conditions.

†GE for fuming H₂SO₄.
‡E for HNO₃ above 80%.
*E for glacial acetic.

Corrodibility of Some Common Metals and Alloys—continued

Class of Material	Carbon	Chromium	Nickel	Silicon	Copper	Other Constituents	Atmospheres Sea Shore	Atmospheres Industrial	Atmospheres Domestic	Water Mine	Water Sea	Water Saline with H₂S	Water Brackish with MCl	Water Wet Steam
Stellite	3·0 max.	30·00				Co 55·0, W 15·0	E	E	E	E	E	E	G	E
Hastelloy A			55·00			Mo 20·0, Fe 20·0	G	G	G	G	G	F	G	G
Hastelloy B			64·00			Mo 28·0, Fe 6·0	G	G	G	G	G	F	G	G
Hastelloy C		14·00	58·00			Mo 17·0, W 5·0, Fe 6·0	E	E	E	E	E	E	F	E
Hastelloy D			85·00	10·0	4·0		G	G	G	E	G	F	E	G
Commercially pure Ni			99·20				E	G	E	F	E	F	F	E
Nickel alloys: Monel metal			67·50	0·50	28·50	Mn 1·50	E	G	E	E	E	E	G	E
Nichrome, 60-15	0·15	15·00	60·00			Fe 6·0	E	G	E	P	E	F	G	E
Inconel, 14% Cr	0·12	14·00	80·00				E	G	E	G	E	E	E	E
80% Ni, 20% Cr		20·00	80·00				E	G	E	G	E	E	E	E
Commercially pure Cu					99·90		G	G	G	G	G	P	F	G
Copper alloys: Red brass					85·00	Zn 15·0	G	G	G	G	G	P	F	G
Tobin bronze					60·00	Zn 39·2, Sn 0·75	G	G	G	G	G	F	G	G
Phosphor bronze					95·00	Sn 5·0	G	G	G	G	G	F	G	G
Silicon bronze				3·00	95·00	Mn 1·0 or Zn 1·5, Sn 0·5	G	G	G	G	G	P	G	G
Aluminium bronze			0·50		90·00	Al 9·0, Fe 0·5	G	G	G	G	G	F	P	G
Nickel silver			20·00		75·00	Zn 5·0	G	G	G	F	G	F	G	G
Admiralty metal					70·00	Zn 29·0, Sn 1·0	G	G	G	G	G	G	G	G
Commercially pure Al-SIC						Al 99·20	G	E	G	G	G	F	F	G
Aluminium alloys: LM4M (DTD 424A)							G	E	F	F	G	P	G	F
LM6M (3L3)							G	E	G	F	G	P	G	G
LM5M or LM10W (L55)							G	E	G	G	G	F	G	G
Commercially pure Mg							F	G	P	P	P	P	P	F
Cast Mg alloys	A.S.T.M. A Z63, A Z92, A Z90						G	G	F	G	P	F	G	G
Wrought Mg alloys	A.S.T.M. M I, AZ31X, A Z61X AZ80X						G	G	G	G	P	F	G	G
Commercially pure Sn							F	G	E	F	P	G	P	P
Commercially pure Pb					0·06	Pb 99·90	G	G	E	G	P	F	P	P

E—Excellent; almost unlimited service.
G—Good; will give good service.
F—Fair.
P—Poor.

§Below 50° C.
¶May crack under certain conditions.

†GE for fuming H₂SO₄.
‡E for HNO₃ above 80%.
*E for glacial acetic.

Corrodibility of Some Common Metals and Alloys—continued

Class of Material	Oxidation Resistance Max. Safe Temp. °F.			Fruit, Vegetable Juices	Dairy Products	Other Food Products	Acids Moderate Concentration 5-15%					Alkalies		Salt Solutions Medium Concentration			Hot Sulphite Liquor	Dye Liquor	Refinery Crudes Below 400° F.	
	Oxidizing Gases	Reducing Fuel Gas	Sulphur Rich Gas				HCl	H₂SO₄	HNO₃	Acetic	Phosphoric	1 to 20% Sol.	Fused	NH₄Cl	MgCl₂	MgSO₄			Sweet	Sour
Ingot iron or wrought iron	1000	1000		P.	P.	P.	P.	P.	P.	P.	P.	G	G	P.	FC	FC	P.		F	P.
Low C steel				P.	P.	P.	P.	P.	P.	P.	P.	G	G	P.	FC	FC	P.	F.	F.	P.
Copper and high tensile steels				P.	P.	P.	P.	P.	P.	P.	P.	E	F	P.	FC	FC	P.	P	F.	P.
Hot galvanized iron and steel				P.	P.	P.	P.	P.	P.	P.	P.	F	L	P.	P.	P.	P.		L.	L.
Calorized iron and steel	1650	1650	1650	E	E	E	P.	E	E	GA.	P.	L	G	GP.	G	G	P.	F	G	G
Grey cast iron	1200	1200		G	G	G	P.	P.	P.	GG	GP.	G	P.	GP.	G	G	P.	P.	G	G
High silicon iron	1750	1750	1750	GF	FG	G	FG	F.	E	GG	F	FG	FG	G	G	G	FF	P.	FG	FG
Nickel cast iron	1400	1400	1200	PF	GG	G	P.	P.	P.	G	FG	E	FG	P.	G	G	P.	P.	G	G
Chromium cast iron				P.	GG	G	FG	FG	G	G	FG	E	FG	FC	F	FC	P.	P.	GG	GP.
Ni-Cr-Cu cast iron				E	GG	G	P.	F	E	GG	F.	F	G	E	G	FC	P.	P.	G	G
Nickel steel: Low Ni				G	GG	GG	P.	FG	P.	G	P.	GE	GG	L	G	L	P.	P.	G	G
High Ni				GF	FG	GG	P.	FG	P.	GG	GP.	GE	FG	E	F	G	P.	P.	G	PF
Chromium steels: 5% Cr	1150	1150	1150	P.	FG	G	FG	FG	G	GG	P.	G	GG	G	G	G	P.		G	G
7% Cr	1250	1250	1200	P.	G	G	P.	F	E	G	G	G	FG	L	G	F	GE	F	E	GE
9% Cr	1300	1350	1250	E	G	G	P.	FG	G	ES	GP.	G	FG	E	G	FC	P.		E	GG
12% Cr	1350	1550	1300	G	G	G	P.	FC	G	G	G	G	G	G	G	G	L.	G	GE	G
17% Cr	1550		1500	GE	GE	GE	FC	F	E	G	G	G	L.	G	G	G	E	G	GE	G
17% Cr, 4% Mo				G	G	G	P.	FG	G	F	F	G	F	G	G	G	E		E	G
27% Cr	1900	1900	1800	E	E	E	P.	FG	E	FC	G	G	G	E	G	G	G	G	E	E
Silcrome (8% Cr, 3% Si)	1500	1500	1100	E	G	G	PF	G	G	G	G	E	F	G	GG	FC	PG	G	GE	GE
Cr-Ni steels: 8-20	1650	1650	1100	E	E	E	P.	G	F	E	G	G	L	G	GG	G	G		E	E
18-8	2100	2000		E	E	E	P.	FC	FC	E	E	E	F	G	G	G	E	G	E	E
18-12	2100	2000		E	E	E	F	FG	FC	E	E	E	L	G	G	G	G	G	E	E
18-35	2100	2000		E	E	E	G	G	E	E	E	E	G	G	G	G	C	G	E	E
25-12	2100	2000	1900	E	E	E	G	PG	G	E	E	E	F	G	GC	G	PG		E	E
25-20	2100	2000		E	E	E		F	FC	E	E	E	G	G	GC	G	G		E	E
Stellite				E	E	E	F	F	FC	E	E	E	C	C	GC	G	G	G	E	E
Hastelloy A	1400	1800	P	C	C	C	C	C	P	E	E	E	P	P	G	G	C	P.	C	C

E—Excellent; almost unlimited service.
G—Good; will give good service.
F—Fair. P—Poor.
†GE for fuming H₂SO₄; ‡E for HNO₃ above 80%; §Below 50° C.; ¶May crack under certain conditions. ‖F for alkali aniline.

22

Corrodibility of Some Common Metals and Alloys—continued

Class of Material	Oxidizing Gases	Reducing Fuel Gas	Sulphur Rich Gas	Fruit, Vegetable Juices	Dairy Products	Other Food Products	HCl	H₂SO₄	HNO₃	Acetic	Phosphoric	Alkalies 1 to 20% Sol.	Alkalies Fused	NH₄Cl	MgCl₂	Na₂SO₄	Hot Sulphite Liquor	Dye Liquor	Refinery Crudes Sweet	Refinery Crudes Sour
Hastelloy B	1400	1800	P	G	G	G	E	E	P	E	E	E	P	E	E	E	P	P	E	E
Hastelloy C	2100	2100	P	G	E	E	E	E	P	E	E	E	P	E	E	E	P	P	E	E
Hastelloy D				E	G	GE	FC	E	GP	E	E	E	P	GE	G	GE	P	P	E	GE
Commercially pure Ni	1900	2300	1000	E	E	FG	FC	FG	P	E	FG	E	P	C	C	G	P	G	E	C
Nickel alloys: Monel metal	1000	2000	1000	G	E	E	FG	G	F	G	C	E	E	E	GE	G	P	E	E	E
Nichrome: 60-15	2050	1900	1500	E	E	E	FG	F	F	G	G	G	G	G	G	G	P	FG	E	E
Inconel: 14% Cr.	2000	2100	1500	FG	P	FC	PG	FG	FG	G	G	G	G	G	G	G	P	FG	FG	PF
80% Ni, 20% Cr	2000	2100		FG	P	FC	PF	FG	FG	FG	G	G	G	G	G	FG	G	FC	G	F
Commercially pure Cu.				PF	P	F	FC	FG	FG	FG	FG	G	F	FG	FG	FG	G	FC	G	F
Copper alloys: Red brass				FG	PF	F	FC	FC	FG	FG	FG	F	PF	FG	FG	FG	C	FC	G	F
Tobin bronze				FG	P	G	FC	FC	FG	FG	FG	E	F	FG	FG	FG	C	FG	G	F
Phosphor bronze				U	G	FG	FC	FC	P	FG	G	F	F	FG	FC	G	C	FC	G	FC
Silicon bronze				FG	U	U	G	FC	G	G	G	P	PF	FG	PF		F	PE	E	FC
Aluminium bronze				G	U	U	P	PE	FG	G	P	E	F	PF	FC		F	PE	F	E
Nickel silver				G	G	G	P	E	P	FG	P	E	F	FC	FG	G	P	F	G	F
Admiralty metal				G	PE	PE				P	P	E	P				P		G	G
Commercially pure Al-SIC	800	800	800	P	PE	PE						E	P				P			
Aluminium alloys: LM4M	P	P	P		P	G						E	P		FG	G	P			
LM6M	300	300	300																	
LM5M or 10W	800	800	800																	
Commercially pure Mg.	800	800																		
Cast Mg alloys	400	400	400																	
Wrought Mg alloys	400	400	400																	
Commercially pure Sn.	400	400	400															G	F	G
Commercially pure Pb.																			G	G

E—Excellent; almost unlimited service.
G—Good; will give good service.
U—Unsuitable.

F—Fair.
P—Poor.

§Below 50° C.
¶May crack under certain conditions.

†GE for fuming H₂SO₄.
‡E for HNO₃ above 80%.
*E for glacial acetic.

23

CENTIMETRES — INCHES

cm.		in.	cm.		in.
2·54	1	0·3937	66·04	26	10·2362
5·08	2	0·7874	68·58	27	10·6299
7·62	3	1·1811	71·12	28	11·0236
10·16	4	1·5748	73·66	29	11·4173
12·70	5	1·9685	76·20	30	11·8110
15·24	6	2·3622	78·74	31	12·2047
17·78	7	2·7559	81·28	32	12·5984
20·32	8	3·1496	83·82	33	12·9921
22·86	9	3·5433	86·36	34	13·3858
25·40	10	3·9370	88·90	35	13·7795
27·94	11	4·3307	91·44	36	14·1732
30·48	12	4·7244	93·98	37	14·5669
33·02	13	5·1181	96·52	38	14·9606
35·56	14	5·5118	99·06	39	15·3543
38·10	15	5·9055	102·60	40	15·7480
40·64	16	6·2992	104·14	41	16·1417
43·18	17	6·6929	106·68	42	16·5354
45·72	18	7·0866	109·22	43	16·9291
48·26	19	7·4803	111·76	44	17·3228
50·80	20	7·8740	114·30	45	17·7165
53·34	21	8·2677	116·84	46	18·1102
55·88	22	8·6614	119·38	47	18·5039
58·42	23	9·0551	121·92	48	18·8976
60·96	24	9·4488	124·46	49	19·2913
63·50	25	9·8425	127·00	50	19·6850

To use this type of table, take the unit which is to be converted from centre column and read conversion direct from left- or right-hand column as the case may be. For example, in above table 10 in. is equal to 25.40 cm. and 10 cm. is equal to 3.9370 in.

CENTIMETRES — INCHES

cm.		in.	cm.		in.
129·54	51	20·0787	193·04	76	29·9213
132·08	52	20·4724	195·58	77	30·3150
134·62	53	20·8661	198·12	78	30·7087
137·16	54	21·2598	200·66	79	31·1024
139·70	55	21·6535	203·20	80	31·4961
142·24	56	22·0472	205·74	81	31·8898
144·78	57	22·4409	208·28	82	32·2835
147·32	58	22·8346	210·82	83	32·6772
149·86	59	23·2283	213·36	84	33·0709
152·40	60	23·6220	215·90	85	33·4646
154·94	61	24·0157	218·44	86	33·8583
157·48	62	24·4094	220·98	87	34·2520
160·02	63	24·8031	223·52	88	34·6457
162·56	64	25·1969	226·06	89	35·0394
165·10	65	25·5906	228·60	90	35·4331
167·64	66	25·9843	231·14	91	35·8268
170·18	67	26·3780	233·68	92	36·2205
172·72	68	26·7717	236·22	93	36·6142
175·26	69	27·1654	238·76	94	37·0079
177·80	70	27·5591	241·30	95	37·4016
180·34	71	27·9528	243·84	96	37·7953
182·88	72	28·3465	246·38	97	38·1890
185·42	73	28·7402	248·92	98	38·5827
187·96	74	29·1339	251·46	99	38·9764
190·50	75	29·5276	254·00	100	39·3701

METRES — FEET

m.		ft.	m.		ft.
0·3048	1	3·28084	7·9248	26	85·302
0·6096	2	6·562	8·2296	27	88·583
0·9144	3	9·843	8·5344	28	91·863
1·2192	4	13·123	8·8392	29	95·144
1·5240	5	16·404	9·1440	30	98·425
1·8288	6	19·685	9·4488	31	101·706
2·1336	7	22·966	9·7536	32	104·987
2·4384	8	26·247	10·0584	33	108·268
2·7432	9	29·528	10·3632	34	111·549
3·0480	10	32·808	10·6680	35	114·829
3·3528	11	36·089	10·9728	36	118·110
3·6576	12	39·370	11·2776	37	121·391
3·9624	13	42·651	11·5824	38	124·672
4·2672	14	45·932	11·8872	39	127·953
4·5720	15	49·213	12·1920	40	131·234
4·8768	16	52·493	12·4968	41	134·514
5·1816	17	55·774	12·8016	42	137·795
5·4864	18	59·055	13·1064	43	141·076
5·7912	19	62·336	13·4112	44	144·357
6·0960	20	65·617	13·7160	45	147·638
6·4008	21	68·898	14·0208	46	150·919
6·7056	22	72·178	14·3256	47	154·199
7·0104	23	75·459	14·6304	48	157·480
7·3152	24	78·740	14·9352	49	160·761
7·6200	25	82·021	15·2400	50	164·042

METRES — FEET

m.		ft.	m.		ft.
15·5448	51	167·323	23·1648	76	249·344
15·8496	52	170·604	23·4696	77	252·625
16·1544	53	173·884	23·7744	78	255·906
16·4592	54	177·165	24·0792	79	259·186
16·7640	55	180·446	24·3840	80	262·467
17·0688	56	183·727	24·6888	81	265·748
17·3736	57	187·008	24·9936	82	269·029
17·6784	58	190·289	25·2984	83	272·310
17·9832	59	193·570	25·6032	84	275·591
18·2880	60	196·850	25·9080	85	278·871
18·5928	61	200·131	26·2128	86	282·152
18·8976	62	203·412	26·5176	87	285·433
19·2024	63	206·693	26·8224	88	288·714
19·5072	64	209·974	27·1272	89	291·995
19·8120	65	213·255	27·4320	90	295·276
20·1168	66	216·535	27·7368	91	298·556
20·4216	67	219·816	28·0416	92	301·837
20·7264	68	223·097	28·3464	93	305·118
21·0312	69	226·378	28·6512	94	308·399
21·3360	70	229·659	28·9560	95	311·680
21·6408	71	232·940	29·2608	96	314·961
21·9456	72	236·220	29·5656	97	318·241
22·2504	73	239·501	29·8704	98	321·522
22·5552	74	242·782	30·1752	99	324·803
22·8600	75	246·063	30·4800	100	328·084

SQ. CENTIMETRES — SQ. INCHES

sq. cm.		sq. in.	sq. cm.		sq. in.
6·452	1	0·155	167·742	26	4·030
12·903	2	0·310	174·193	27	4·185
19·355	3	0·465	180·645	28	4·340
25·806	4	0·620	187·096	29	4·495
32·258	5	0·775	193·548	30	4·650
38·710	6	0·930	200·000	31	4·805
45·161	7	1·085	206·451	32	4·960
51·613	8	1·240	212·903	33	5·115
58·064	9	1·395	219·354	34	5·270
64·516	10	1·550	225·806	35	5·425
70·968	11	1·705	232·258	36	5·580
77·419	12	1·860	238·709	37	5·735
83·871	13	2·015	245·161	38	5·890
90·322	14	2·170	251·612	39	6·045
96·774	15	2·325	258·064	40	6·200
103·226	16	2·480	264·516	41	6·355
109·677	17	2·635	270·967	42	6·510
116·129	18	2·790	277·419	43	6·665
122·580	19	2·945	283·870	44	6·820
129·032	20	3·100	290·322	45	6·975
135·484	21	3·255	296·774	46	7·130
141·935	22	3·410	303·225	47	7·285
148·387	23	3·565	309·677	48	7·440
154·838	24	3·720	316·128	49	7·595
161·290	25	3·875	322·580	50	7·750

SQ. CENTIMETRES — SQ. INCHES

sq. cm.		sq. in.	sq. cm.		sq. in.
329·032	51	7·905	490·322	76	11·780
335·483	52	8·060	496·773	77	11·935
341·935	53	8·215	503·225	78	12·090
348·386	54	8·370	509·676	79	12·245
354·838	55	8·525	516·128	80	12·400
361·290	56	8·680	522·579	81	12·555
367·741	57	8·835	529·031	82	12·710
374·193	58	8·990	535·483	83	12·865
380·644	59	9·145	541·934	84	13·020
387·096	60	9·300	548·386	85	13·175
393·548	61	9·455	554·838	86	13·330
399·999	62	9·610	561·289	87	13·485
406·451	63	9·765	567·741	88	13·640
412·902	64	9·920	574·192	89	13·795
419·354	65	10·075	580·644	90	13·950
425·806	66	10·230	587·096	91	14·105
432·257	67	10·385	593·547	92	14·260
438·709	68	10·540	599·999	93	14·415
445·160	69	10·695	606·450	94	14·570
451·612	70	10·850	612·902	95	14·725
458·064	71	11·005	619·354	96	14·880
464·515	72	11·160	625·805	97	15·035
470·967	73	11·315	632·257	98	15·190
477·418	74	11·470	638·708	99	15·345
483·878	75	11·625	645·160	100	15·500

CUBIC CENTIMETRES — CUBIC INCHES

cu. cm.		cu. in.	cu. cm.		cu. in.
16·387064	1	0·061024	426·064	26	1·5866
32·774	2	0·1220	442·451	27	1·6476
49·161	3	0·1831	458·838	28	1·7087
65·548	4	0·2441	475·225	29	1·7697
81·935	5	0·3051	491·612	30	1·8307
98·322	6	0·3661	507·999	31	1·8917
114·709	7	0·4272	524·386	32	1·9528
131·097	8	0·4882	540·773	33	2·0138
147·484	9	0·5492	557·160	34	2·0748
163·871	10	0·6102	573·547	35	2·1358
180·258	11	0·6713	589·934	36	2·1969
196·645	12	0·7323	606·321	37	2·2579
213·032	13	0·7933	622·708	38	2·3189
229·419	14	0·8543	639·096	39	2·3794
245·806	15	0·9154	655·483	40	2·4410
262·193	16	0·9764	671·876	41	2·5020
278·580	17	1·0374	688·257	42	2·5630
294·967	18	1·0984	704·644	43	2·6240
311·354	19	1·1595	721·031	44	2·6851
327·741	20	1·2205	737·418	45	2·7461
344·128	21	1·2815	753·805	46	2·8071
360·515	22	1·3425	770·192	47	2·8681
376·902	23	1·4036	786·579	48	2·9292
393·290	24	1·4646	802·966	49	2·9902
409·677	25	1·5256	819·353	50	3·0512

CUBIC CENTIMETRES — CUBIC INCHES

cu. cm.		cu. in.	cu. cm.		cu. in.
835·740	51	3·1122	1245·417	76	4·6378
852·127	52	3·1732	1261·804	77	4·6988
868·514	53	3·2343	1278·191	78	4·7599
884·901	54	3·2953	1294·578	79	4·8209
901·288	55	3·3563	1310·965	80	4·8819
917·675	56	3·4173	1327·352	81	4·9429
934·062	57	3·4784	1343·739	82	5·0040
950·449	58	3·5394	1360·126	83	5·0650
966·837	59	3·6004	1376·513	84	5·1260
983·224	60	3·6614	1392·900	85	5·1870
999·611	61	3·7225	1409·288	86	5·2481
1015·998	62	3·7835	1425·675	87	5·3091
1032·385	63	3·8445	1442·062	88	5·3701
1048·772	64	3·9055	1458·449	89	5·4311
1065·159	65	3·9666	1474·836	90	5·4922
1081·546	66	4·0276	1491·223	91	5·5532
1097·933	67	4·0886	1507·610	92	5·6142
1114·320	68	4·1496	1523·997	93	5·6752
1130·707	69	4·2107	1540·384	94	5·7363
1147·094	70	4·2717	1556·771	95	5·7973
1163·481	71	4·3327	1573·158	96	5·8583
1179·868	72	4·3937	1589·545	97	5·9193
1196·255	73	4·4548	1605·932	98	5·9804
1212·642	74	4·5158	1622·319	99	6·0414
1229·295	75	4·5768	1638·706	100	6·1024

GRAMMES — OUNCES (Avoirdupois)

grammes		oz.	grammes		oz.
28·3495	1	0·03527	737·0870	26	0·91702
56·6990	2	0·07054	765·4365	27	0·95229
85·0485	3	0·10581	793·7860	28	0·98756
113·3980	4	0·14108	822·1355	29	1·02283
141·7475	5	0·17635	850·4850	30	1·05810
170·0970	6	0·21162	878·8345	31	1·09337
198·4465	7	0·24689	907·1840	32	1·12864
226·7960	8	0·28216	935·5335	33	1·16391
255·1455	9	0·31743	963·8830	34	1·19918
283·4950	10	0·35270	992·2325	35	1·23445
311·8445	11	0·38797	1020·5820	36	1·26972
340·1940	12	0·42324	1048·9315	37	1·30499
368·5435	13	0·45851	1077·2810	38	1·34026
396·8930	14	0·49378	1105·6305	39	1·37553
425·2425	15	0·52905	1133·9800	40	1·41080
453·5920	16	0·56432	1162·3295	41	1·44607
481·9415	17	0·59959	1190·6790	42	1·48134
510·2910	18	0·63486	1219·0285	43	1·51661
538·6405	19	0·67013	1247·3780	44	1·55188
566·9900	20	0·70540	1275·7275	45	1·58715
595·3395	21	0·74067	1304·0770	46	1·62242
623·6890	22	0·77594	1332·4265	47	1·65769
652·0385	23	0·81121	1360·7760	48	1·69296
680·3880	24	0·84648	1389·1255	49	1·72823
708·7375	25	0·88175	1417·4750	50	1·76350

GRAMMES — OUNCES (Avoirdupois)

grammes		oz.	grammes		oz.
1445·8245	51	1·79877	2154·5620	76	2·68052
1474·1740	52	1·83404	2182·9115	77	2·71579
1502·5235	53	1·86931	2211·2610	78	2·75106
1530·8730	54	1·90458	2239·6105	79	2·78633
1559·2225	55	1·93985	2267·9600	80	2·82160
1587·5720	56	1·97512	2296·3095	81	2·85687
1615·9215	57	2·01039	2324·6590	82	2·89214
1644·2710	58	2·04566	2353·0085	83	2·92741
1672·6205	59	2·08093	2381·3580	84	2·96268
1700·9700	60	2·11620	2409·7075	85	2·99795
1729·3195	61	2·15147	2438·0570	86	3·03322
1757·6690	62	2·18674	2466·4065	87	3·06849
1786·0185	63	2·22201	2494·7560	88	3·10376
1814·3680	64	2·25728	2523·1055	89	3·13903
1842·7175	65	2·29255	2551·4550	90	3·17430
1871·0670	66	2·32782	2579·8045	91	3·20957
1899·4165	67	2·36309	2608·1540	92	3·24484
1927·7660	68	2·39836	2636·5035	93	3·28011
1956·1155	69	2·43363	2664·8530	94	3·31538
1984·4650	70	2·46890	2693·2025	95	3·35065
2012·8145	71	2·50417	2721·5520	96	3·38592
2041·1640	72	2·53944	2749·9015	97	3·42119
2069·5135	73	2·57471	2778·2510	98	3·45646
2097·8630	74	2·60998	2806·6005	99	3·49173
2126·2125	75	2·64525	2834·9500	100	3·52700

KILOGRAMMES — POUNDS

kg.		lb.	kg.		lb.
0·45359243	1	2·20462	11·793	26	57·320
0·907	2	4·409	12·247	27	59·525
1·361	3	6·614	12·701	28	61·729
1·814	4	8·818	13·154	29	63·934
2·268	5	11·023	13·608	30	66·139
2·722	6	13·228	14·061	31	68·343
3·175	7	15·432	14·515	32	70·548
3·629	8	17·637	14·969	33	72·753
4·082	9	19·842	15·422	34	74·957
4·536	10	22·046	15·876	35	77·162
4·990	11	24·251	16·329	36	79·366
5·443	12	26·455	16·783	37	81·571
5·897	13	28·660	17·237	38	83·776
6·350	14	30·865	17·690	39	85·980
6·804	15	33·069	18·144	40	88·185
7·257	16	35·274	18·597	41	90·390
7·711	17	37·479	19·051	42	92·594
8·165	18	39·683	19·504	43	94·799
8·618	19	41·888	19·958	44	97·003
9·072	20	44·092	20·412	45	99·208
9·525	21	46·297	20·865	46	101·413
9·979	22	48·502	21·319	47	103·617
10·433	23	50·706	21·772	48	105·822
10·886	24	52·911	22·226	49	108·026
11·340	25	55·116	22·680	50	110·231

KILOGRAMMES — POUNDS

kg.		lb.	kg.		lb.
23·133	51	112·436	34·473	76	167·551
23·587	52	114·640	34·927	77	169·756
24·040	53	116·845	35·380	78	171·961
24·494	54	119·050	35·834	79	174·165
24·948	55	121·254	36·287	80	176·370
25·401	56	123·459	36·741	81	178·574
25·855	57	125·663	37·195	82	180·779
26·308	58	127·868	37·648	83	182·984
26·762	59	130·073	38·102	84	185·188
27·216	60	132·277	38·555	85	187·393
27·669	61	134·482	39·009	86	189·598
28·123	62	136·687	39·463	87	191·802
28·576	63	138·891	39·916	88	194·007
29·030	64	141·096	40·370	89	196·211
29·484	65	143·300	40·823	90	198·416
29·937	66	145·505	41·277	91	200·621
30·391	67	147·710	41·731	92	202·825
30·844	68	149·914	42·184	93	205·030
31·298	69	152·119	42·638	94	207·235
31·751	70	154·324	43·091	95	209·439
32·205	71	156·528	43·545	96	211·644
32·659	72	158·733	43·999	97	213·848
33·112	73	160·937	44·452	98	216·053
33·566	74	163·142	44·906	99	218·258
34·019	75	165·347	45·359	100	220·462

LONG TONS (2240 lb.) — SHORT TONS (2000 lb.)

long tons		short tons	long tons		short tons
0·89286	1	1·12	23·2143	26	29·12
1·7857	2	2·24	24·1071	27	30·24
2·6786	3	3·36	25·0000	28	31·36
3·5714	4	4·48	25·8929	29	32·48
4·4643	5	5·60	26·7857	30	33·60
5·3571	6	6·72	27·6786	31	34·72
6·2500	7	7·84	28·5714	32	35·84
7·1429	8	8·96	29·4643	33	36·96
8·0357	9	10·08	30·3571	34	38·08
8·9286	10	11·20	31·2500	35	39·20
9·8214	11	12·32	32·1429	36	40·32
10·7143	12	13·44	33·0357	37	41·44
11·6071	13	14·56	33·9286	38	42·56
12·500	14	15·68	34·8214	39	43·68
13·3929	15	16·80	35·7143	40	44·80
14·2857	16	17·92	36·6071	41	45·92
15·1786	17	19·04	37·5000	42	47·04
16·0714	18	20·16	38·3929	43	48·16
16·9643	19	21·28	39·2857	44	49·28
17·8571	20	22·40	40·1786	45	50·40
18·7500	21	23·52	41·0714	46	51·52
19·6429	22	24·64	41·9643	47	52·64
20·5357	23	25·76	42·8571	48	53·76
21·4286	24	26·88	43·7500	49	54·88
22·3214	25	28·00	44·6429	50	56·00

LONG TONS (2240 lb.) — SHORT TONS (2000 lb.)

long tons		short tons	long tons		short tons
45·5357	51	57·12	67·8571	76	85·12
46·4286	52	58·24	68·7500	77	86·24
47·3214	53	59·36	69·6429	78	87·36
48·2143	54	60·48	70·5357	79	88·48
49·1071	55	61·60	71·4286	80	89·60
50·0000	56	62·72	72·3214	81	90·72
50·8929	57	63·84	73·2143	82	91·84
51·7857	58	64·96	74·1071	83	92·96
52·6786	59	66·08	75·0000	84	94·08
53·5714	60	67·20	75·8929	85	95·20
54·4643	61	68·32	76·7857	86	96·32
55·3571	62	69·44	77·6786	87	97·44
56·2500	63	70·56	78·5714	88	98·56
57·1429	64	71·68	79·4643	89	99·68
58·0357	65	72·80	80·3571	90	100·80
58·9286	66	73·92	81·2500	91	101·92
59·8214	67	75·04	82·1429	92	103·04
60·7143	68	76·16	83·0357	93	104·16
61·6071	69	77·28	83·9286	94	105·28
62·5000	70	78·40	84·8214	95	106·40
63·3929	71	79·52	85·7143	96	107·52
64·2857	72	80·64	86·6071	97	108·64
65·1786	73	81·76	87·5000	98	109·76
66·0714	74	82·88	88·3929	99	110·88
66·9643	75	84·00	89·2857	100	112·00

METRIC TONNES (1000 kg.) — TONS (2240 lb.)

tonnes		long tons	tonnes		long tons
1·01605	1	0·984206	26·417	26	25·589
2·032	2	1·968	27·433	27	26·574
3·048	3	2·953	28·449	28	27·558
4·064	4	3·937	29·465	29	28·542
5·080	5	4·921	30·481	30	29·526
6·096	6	5·905	31·497	31	30·510
7·112	7	6·889	32·514	32	31·495
8·128	8	7·874	33·530	33	32·479
9·144	9	8·858	34·546	34	33·463
10·160	10	9·842	35·562	35	34·447
11·177	11	10·826	36·578	36	35·431
12·193	12	11·810	37·594	37	36·416
13·209	13	12·795	38·610	38	37·400
14·225	14	13·779	39·626	39	38·384
15·241	15	14·763	40·642	40	39·368
16·257	16	15·747	41·658	41	40·352
17·273	17	16·732	42·674	42	41·337
18·289	18	17·716	43·690	43	42·321
19·305	19	18·700	44·706	44	43·305
20·321	20	19·684	45·722	45	44·289
21·337	21	20·668	46·738	46	45·273
22·353	22	21·653	47·754	47	46·258
23·369	23	22·637	48·770	48	47·242
24·385	24	23·621	49·786	49	48·226
25·401	25	24·605	50·802	50	49·210

METRIC TONNES (1000 kg.) — TONS (2240 lb.)

tonnes		long tons	tonnes		long tons
51·818	51	50·195	77·220	76	74·800
52·834	52	51·179	78·236	77	75·784
53·850	53	52·163	79·252	78	76·768
54·867	54	53·147	80·268	79	77·752
55·883	55	54·131	81·284	80	78·736
56·899	56	55·116	82·300	81	79·721
57·915	57	56·100	83·316	82	80·705
58·931	58	57·084	84·332	83	81·689
59·947	59	58·068	85·348	84	82·673
60·963	60	59·052	86·364	85	83·658
61·979	61	60·037	87·380	86	84·642
62·995	62	61·021	88·396	87	85·626
64·011	63	62·005	89·412	88	86·610
65·027	64	62·989	90·428	89	87·594
66·043	65	63·973	91·444	90	88·579
67·059	66	64·958	92·460	91	89·563
68·075	67	65·942	93·476	92	90·547
69·091	68	66·926	94·492	93	91·531
70·107	69	67·910	95·508	94	92·515
71·123	70	68·894	96·524	95	93·500
72·139	71	69·879	97·541	96	94·484
73·155	72	70·863	98·557	97	95·468
74·171	73	71·847	99·573	98	96·452
75·187	74	72·831	100·589	99	97·436
76·204	75	73·815	101·605	100	98·421

METRIC TONNES (1000 kg.) — SHORT TONS (2000 lb.)

metric tonnes		short tons	metric tonnes		short tons
0·90718	1	1·1023	23·5868	26	28·6601
1·8144	2	2·2046	24·4940	27	29·7624
2·7216	3	3·3069	25·4012	28	30·8647
3·6287	4	4·4092	26·3084	29	31·9669
4·5359	5	5·5115	27·2156	30	33·0693
5·4431	6	6·6139	28·1228	31	34·1716
6·3503	7	7·7162	29·0299	32	35·2739
7·2575	8	8·8185	29·9371	33	36·3762
8·1647	9	9·9208	30·8443	34	37·4785
9·0719	10	11·0231	31·7515	35	38·5809
9·9791	11	12·1254	32·6587	36	39·6832
10·8863	12	13·2277	33·5658	37	40·7855
11·7934	13	14·3300	34·4730	38	41·8878
12·7006	14	15·4323	35·3802	39	42·9901
13·6078	15	16·5347	36·2874	40	44·0924
14·5150	16	17·6370	37·1946	41	45·1947
15·4222	17	18·7393	38·1018	42	46·2970
16·3294	18	19·8416	39·0089	43	47·3993
17·2365	19	20·9439	39·9161	44	48·5016
18·1437	20	22·0462	40·8233	45	49·6040
19·0501	21	23·1485	41·7305	46	50·7063
19·9581	22	24·2508	42·6377	47	51·8086
20·8653	23	25·3531	43·5449	48	52·9109
21·7724	24	26·4554	44·4521	49	54·0132
22·6796	25	27·5578	45·3592	50	55·1155

METRIC TONNES (1000 kg.) — SHORT TONS (2000 lb.)

metric tonnes		short tons	metric tonnes		short tons
46·2664	51	56·2178	68·9460	76	83·7756
47·1736	52	57·3201	69·8532	77	84·8779
48·0808	53	58·4224	70·7604	78	85·9802
48·9880	54	59·5247	71·6676	79	87·0825
49·8952	55	60·6271	72·5748	80	88·1848
50·8024	56	61·7294	73·4820	81	89·2871
51·7095	57	62·8317	74·3891	82	90·3894
52·6167	58	63·9340	75·2963	83	91·4917
53·5239	59	65·0363	76·2035	84	92·5940
54·4311	60	66·1386	77·1107	85	93·6964
55·3383	61	67·2409	78·0179	86	94·7987
56·2454	62	68·3432	78·9251	87	95·9010
57·1526	63	69·4455	79·8323	88	97·0033
58·0596	64	70·5478	80·7395	89	98·1056
58·9670	65	71·6502	81·6466	90	99·2079
59·8742	66	72·7525	82·5538	91	100·3102
60·7814	67	73·8548	83·4610	92	101·4125
61·6886	68	74·9571	84·3682	93	102·5148
62·5958	69	76·0594	85·2754	94	103·6171
63·5029	70	77·1612	86·1826	95	104·7195
64·4101	71	78·2640	87·0897	96	105·8218
65·3173	72	79·3663	87·9969	97	106·9241
66·2245	73	80·4686	88·9041	98	108·0264
67·1317	74	81·5709	89·8113	99	109·1287
68·0389	75	82·6733	90·7185	100	110·2310

CONVERSION TABLES

KILOGRAMMES INTO LONG TONS, CWTS., ETC.

Kilos	Qrs.	Lbs.	Kilos	Cwts.	Qrs.	Lbs.
1		2·2	60	1	0	20·3
2		4·4	70	1	1	14·3
3		6·6	80	1	2	8·4
4		8·8	90	1	3	2·4
5		11·0	100	1	3	24·5
6		13·2	200	3	3	20·9
7		15·4	300	5	3	17·4
8		17·6	400	7	3	13·8
9		19·8	500	9	3	10·3
10		22·0	600	11	3	6·8
20	1	16·1	700	13	3	3·2
30	2	10·4	800	15	2	27·7
40	3	4·2	900	17	2	24·1
50	3	26·2	1000	19	2	20·6

Kilos	Tons	Cwts.	Qrs.	Lbs.
2000	1	19	1	13·2
3000	2	19	0	5·8
4000	3	18	2	26·4
5000	4	18	1	19·0
6000	5	18	0	11·6
7000	6	17	3	4·2
8000	7	17	1	24·8
9000	8	17	0	17·4
10000	9	16	3	10·0
15000	14	15	1	1·0
20000	19	13	2	20·0
25000	24	12	0	11·0
50000	49	4	0	22·0

KILOGRAMMES INTO SHORT TONS (2,000 lb.), CWTS. (100 lb.) AND LB.

Kilos	Cwt.	Lbs.	Kilos	Cwt.	Lbs.
1	—	2·2	60	1	32·3
2	—	4·4	70	1	54·3
3	—	6·6	80	1	76·4
4	—	8·8	90	1	98·4
5	—	11·0	100	2	20·5
6	—	13·2	200	4	40·9
7	—	15·4	300	6	61·4
8	—	17·6	400	8	81·8
9	—	19·8	500	11	2·3
10	—	22·0	600	13	22·8
20	—	44·1	700	15	43·2
30	—	66·1	800	17	63·7
40	—	88·2	900	19	84·1
50	1	10·2			

Kilos	Short Tons	Cwts.	Lbs.
1000	1	2	4·6
2000	2	4	9·2
3000	3	6	13·9
4000	4	8	18·5
5000	5	10	23·1
6000	6	12	27·7
7000	7	14	32·3
8000	8	16	37·0
9000	9	18	41·6
10000	11	0	46·2
15000	16	10	69·3
20000	22	0	92·4
25000	27	11	15·5
50000	55	2	31·0

LITRES — IMPERIAL GALLONS

litres		Imp. gall.	litres		Imp. gall.
4·5459631	1	0·219975	118·195	26	5·7194
9·092	2	0·4400	122·741	27	5·9393
13·638	3	0·6599	127·287	28	6·1593
18·184	4	0·8799	131·833	29	6·3793
22·730	5	1·0999	136·379	30	6·5993
27·276	6	1·3199	140·925	31	6·8192
31·822	7	1·5398	145·471	32	7·0392
36·368	8	1·7598	150·017	33	7·2592
40·914	9	1·9798	154·563	34	7·4792
45·460	10	2·1998	159·109	35	7·6991
50·006	11	2·4197	163·655	36	7·9191
54·552	12	2·6397	168·201	37	8·1391
59·098	13	2·8597	172·747	38	8·3591
63·643	14	3·0797	177·293	39	8·5790
68·189	15	3·2996	181·839	40	8·7990
72·735	16	3·5196	186·384	41	9·0190
77·281	17	3·7396	190·930	42	9·2390
81·827	18	3·9596	195·476	43	9·4589
86·373	19	4·1795	200·022	44	9·6789
90·919	20	4·3995	204·568	45	9·8989
95·465	21	4·6195	209·114	46	10·1189
100·011	22	4·8395	213·660	47	10·3388
104·557	23	5·0594	218·206	48	10·5588
109·103	24	5·2794	222·752	49	10·7788
113·649	25	5·4994	227·298	50	10·9988

LITRES — IMPERIAL GALLONS

litres		Imp. gall.	litres		Imp. gall.
231·844	51	11·2187	345·493	76	16·7181
236·390	52	11·4387	350·039	77	16·9381
240·936	53	11·6587	354·585	78	17·1581
245·482	54	11·8787	359·131	79	17·3780
250·028	55	12·0986	363·677	80	17·5980
254·574	56	12·3186	368·223	81	17·8180
259·120	57	12·5386	372·769	82	18·0380
263·666	58	12·7586	377·315	83	18·2579
268·212	59	12·9785	381·861	84	18·4779
272·758	60	13·1985	386·407	85	18·6979
277·304	61	13·4185	390·953	86	18·9179
281·850	62	13·6385	395·499	87	19·1379
286·396	63	13·8584	400·045	88	19·3578
290·942	64	14·0784	404·591	89	19·5778
295·488	65	14·2984	409·137	90	19·7978
300·034	66	14·5184	413·683	91	20·0178
304·580	67	14·7383	418·229	92	20·2377
309·125	68	14·9583	422·775	93	20·4577
313·671	69	15·1783	427·321	94	20·6777
318·217	70	15·3983	431·866	95	20·8977
322·763	71	15·6182	436·412	96	21·1176
327·309	72	15·8382	440·958	97	21·3376
331·855	73	16·0582	445·504	98	21·5576
336·401	74	16·2782	450·050	99	21·7776
340·947	75	16·4981	454·596	100	21·9975

U.S. GALLONS — IMPERIAL GALLONS

U.S. gall.		Imp. gall.	U.S. gall.		Imp. gall.
1·20095	1	0·83267	31·225	26	21·649
2·402	2	1·665	32·426	27	22·482
3·603	3	2·498	33·627	28	23·315
4·804	4	3·331	34·828	29	24·147
6·005	5	4·163	36·029	30	24·980
7·206	6	4·996	37·229	31	25·813
8·407	7	5·829	38·430	32	26·645
9·608	8	6·661	39·631	33	27·478
10·809	9	7·494	40·832	34	28·311
12·010	10	8·327	42·033	35	29·143
13·210	11	9·159	43·234	36	29·976
14·411	12	9·992	44·435	37	30·809
15·612	13	10·825	45·636	38	31·641
16·813	14	11·657	46·837	39	32·474
18·014	15	12·490	48·038	40	33·307
19·215	16	13·323	49·239	41	34·139
20·416	17	14·155	50·440	42	34·972
21·617	18	14·988	51·641	43	35·805
22·818	19	15·821	52·842	44	36·637
24·019	20	16·653	54·043	45	37·470
25·220	21	17·486	55·244	46	38·303
26·421	22	18·319	56·445	47	39·135
27·622	23	19·151	57·646	48	39·968
28·823	24	19·984	58·847	49	40·801
30·024	25	20·817	60·047	50	41·634

U.S. GALLONS — IMPERIAL GALLONS

U.S. gall.		Imp.gall.	U.S. gall.		Imp. gall.
61·248	51	42·466	91·272	76	63·283
62·449	52	43·299	92·473	77	64·116
63·650	53	44·132	93·674	78	64·948
64·851	54	44·964	94·875	79	65·781
66·052	55	45·797	96·076	80	66·614
67·253	56	46·630	97·277	81	67·446
68·454	57	47·462	98·478	82	68·279
69·655	58	48·295	99·679	83	69·112
70·856	59	49·128	100·880	84	69·944
72·057	60	49·960	102·081	85	70·777
73·258	61	50·793	103·282	86	71·610
74·459	62	51·626	104·483	87	72·442
75·660	63	52·458	105·684	88	73·275
76·861	64	53·291	106·885	89	74·108
78·062	65	54·124	108·086	90	74·940
79·263	66	54·956	109·286	91	75·773
80·464	67	55·789	110·487	92	76·606
81·665	68	56·622	111·688	93	77·438
82·866	69	57·454	112·889	94	78·271
84·067	70	58·287	114·090	95	79·104
85·267	71	59·120	115·291	96	79·936
86·468	72	59·952	116·492	97	80·769
87·669	73	60·785	117·693	98	81·602
88·870	74	61·618	118·894	99	82·434
90·071	75	62·450	120·095	100	83·267

LITRES — U.S. GALLONS

litres		U.S. gall.	litres		U.S. gall.
3·78530	1	0·26418	98·418	26	6·869
7·571	2	0·5284	102·203	27	7·133
11·356	3	0·7925	105·988	28	7·397
15·141	4	1·0567	109·774	29	7·661
18·926	5	1·3209	113·559	30	7·926
22·712	6	1·5851	117·344	31	8·190
26·497	7	1·8493	121·129	32	8·454
30·282	8	2·1134	124·915	33	8·718
34·068	9	2·3776	128·700	34	8·982
37·853	10	2·6418	132·485	35	9·246
41·638	11	2·9060	136·271	36	9·510
45·424	12	3·1702	140·056	37	9·775
49·209	13	3·4343	143·842	38	10·039
52·994	14	3·6985	147·627	39	10·303
56·780	15	3·9627	151·412	40	10·567
60·565	16	4·2265	155·197	41	10·831
64·350	17	4·4911	158·986	42	11·096
68·135	18	4·7552	162·768	43	11·360
71·921	19	5·0194	166·553	44	11·624
75·706	20	5·2835	170·339	45	11·888
79·491	21	5·5478	174·124	46	12·152
83·277	22	5·8120	177·909	47	12·416
87·062	23	6·0761	181·695	48	12·681
90·847	24	6·3403	185·480	49	12·945
94·633	25	6·6045	189·265	50	13·209

LITRES — U.S. GALLONS

litres		U.S. gall.	litres		U.S. gall.
193·050	51	13·473	287·683	76	20·078
196·836	52	13·737	291·468	77	20·342
200·621	53	14·002	295·254	78	20·606
204·406	54	14·266	299·039	79	20·870
208·192	55	14·530	302·824	80	21·134
211·977	56	14·794	306·610	81	21·399
215·763	57	15·058	310·395	82	21·662
219·548	58	15·322	314·181	83	21·927
223·332	59	15·587	317·965	84	22·191
227·118	60	15·851	321·751	85	22·455
230·903	61	16·115	325·536	86	22·719
234·689	62	16·379	329·321	87	22·984
238·474	63	16·643	333·107	88	23·248
242·259	64	16·907	336·892	89	23·512
246·044	65	17·172	340·677	90	23·776
249·830	66	17·436	344·463	91	24·040
253·615	67	17·700	348·248	92	24·305
257·401	68	17·964	352·033	93	24·569
261·186	69	18·228	355·818	94	24·833
264·971	70	18·493	359·604	95	25·097
268·757	71	18·757	363·389	96	25·361
272·542	72	19·021	367·174	97	25·625
276·327	73	19·285	370·960	98	25·890
280·112	74	19·549	374·745	99	26·154
283·898	75	19·813	378·530	100	26·418

KG./SQ. MM. — LONG TONS/SQ. IN.

kg./sq.mm.		tons/sq. in.	kg/sq.mm.		tons/sq. in.
1·57488	1	0·63497	40·95	26	16·51
3·15	2	1·2700	42·52	27	17·14
4·72	3	1·90	44·10	28	17·78
6·30	4	2·54	45·67	29	18·41
7·87	5	3·17	47·25	30	19·05
9·45	6	3·81	48·82	31	19·68
11·02	7	4·44	50·40	32	20·32
12·60	8	5·08	51·97	33	20·95
14·17	9	5·71	53·55	34	21·59
15·75	10	6·35	55·12	35	22·22
17·32	11	6·98	56·70	36	22·86
18·90	12	7·62	58·27	37	23·49
20·47	13	8·25	59·85	38	24·13
22·05	14	8·89	61·42	39	24·76
23·62	15	9·52	63·00	40	25·40
25·20	16	10·16	64·57	41	26·03
26·77	17	10·79	66·14	42	26·67
28·35	18	11·43	67·72	43	27·30
29·92	19	12·06	69·29	44	27·94
31·50	20	12·70	70·87	45	28·57
33·07	21	13·33	72·44	46	29·21
34·65	22	13·97	74·02	47	29·84
36·22	23	14·60	75·59	48	30·48
37·80	24	15·24	77·17	49	31·11
39·37	25	15·87	78·74	50	31·75

KG./SQ. MM. — LONG TONS/SQ. IN.

kg./sq.mm.		tons/sq. in.	kg./sq.mm.		tons/sq. in.
80·32	51	32·38	119·69	76	48·26
81·89	52	33·02	121·27	77	48·89
83·47	53	33·65	122·84	78	49·53
85·04	54	34·29	124·42	79	50·16
86·62	55	34·92	125·99	80	50·80
88·19	56	35·56	127·57	81	51·43
89·77	57	36·19	129·14	82	52·07
91·34	58	36·83	130·72	83	52·70
92·92	59	37·46	132·29	84	53·34
94·49	60	38·10	133·86	85	53·97
96·07	61	38·73	135·44	86	54·61
97·64	62	39·37	137·01	87	55·24
99·22	63	40·00	138·59	88	55·88
100·79	64	40·64	140·16	89	56·51
102·37	65	41·27	141·74	90	57·15
103·94	66	41·91	143·31	91	57·78
105·52	67	42·54	144·89	92	58·42
107·09	68	43·18	146·46	93	59·05
108·67	69	43·81	148·04	94	59·69
110·24	70	44·45	149·61	95	60·32
111·82	71	45·08	151·19	96	60·96
113·39	72	45·72	152·76	97	61·59
114·97	73	46·35	154·34	98	62·23
116·54	74	46·99	155·91	99	62·86
118·12	75	47·62	157·49	100	63·50

KG./SQ. MM. — SHORT TONS/SQ. IN.

kg./sq.mm.		short tons/sq. in.	kg./sq.mm.		short tons/sq. in.
1·41	1	0·71	36·56	26	18·45
2·81	2	1·42	37·96	27	19·16
4·22	3	2·13	39·37	28	19·87
5·62	4	2·84	40·78	29	20·58
7·03	5	3·55	42·18	30	21·29
8·44	6	4·26	43·59	31	22·00
9·84	7	4·97	44·99	32	22·71
11·25	8	5·68	46·40	33	23·42
12·65	9	6·39	41·81	34	24·13
14·06	10	7·10	49·21	35	24·84
15·47	11	7·81	50·62	36	25·55
16·87	12	8·52	52·02	37	26·26
18·28	13	9·23	53·43	38	26·97
19·68	14	9·94	54·84	39	27·68
21·09	15	10·65	56·24	40	28·39
22·50	16	11·35	57·65	41	29·10
23·90	17	12·06	59·05	42	29·8!
25·31	18	12·77	60·46	43	30·52
26·71	19	13·83	61·87	44	31·23
28·12	20	14·19	63·27	45	31·94
29·53	21	14·90	64·68	46	32·64
30·93	22	15·61	66·08	47	33·35
32·34	23	16·32	67·49	48	34·06
33·75	24	17·03	68·90	49	34·77
35·15	25	17·74	70·30	50	35·48

KG./SQ. MM. — SHORT TONS/SQ. IN.

kg./sq.mm.		short tons/sq. in.	kg./sq.mm.		short tons/sq. in.
71·71	51	36·19	106·86	76	53·94
73·11	52	36·90	108·27	77	54·64
74·52	53	37·61	109·67	78	55·35
75·93	54	38·32	111·08	79	56·06
77·33	55	39·03	112·48	80	56·77
78·74	56	39·74	113·89	81	57·48
80·14	57	40·45	115·30	82	58·19
81·55	58	41·16	116·70	83	58·90
82·96	59	41·87	118·11	84	59·61
84·36	60	42·58	119·51	85	60·32
85·77	61	43·29	121·92	86	61·03
87·18	62	44·00	122·33	87	61·74
88·58	63	44·71	123·73	88	62·45
89·99	64	45·42	125·14	89	63·16
91·39	65	46·13	126·54	90	63·87
92·80	66	46·84	127·95	91	64·58
94·21	67	47·55	129·36	92	65·29
95·61	68	48·26	130·76	93	66·00
97·02	69	48·97	132·17	94	66·71
98·42	70	49·68	133·57	95	67·42
99·83	71	50·39	134·98	96	68·13
101·24	72	51·10	136·39	97	68·84
102·64	73	51·81	137·79	98	69·55
104·05	74	52·52	139·20	99	70·26
105·45	75	53·23	140·60	100	70·97

KG./SQ. CM. — LB./SQ. IN.

kg./sq. cm.		lb./sq. in.	kg./sq. cm.		lb/sq. in.
0·0703	1	14·2233	1·8280	26	369·81
0·1406	2	28·45	1·8983	27	384·03
0·2109	3	42·67	1·9686	28	398·25
0·2812	4	56·89	2·0389	29	412·48
0·3515	5	71·12	2·1092	30	426·70
0·4218	6	85·34	2·1795	31	440·92
0·4921	7	99·56	2·2498	32	455·15
0·5625	8	113·79	2·3201	33	469·37
0·6328	9	128·01	2·3904	34	483·59
0·7031	10	142·23	2·4607	35	497·82
0·7734	11	156·46	2·5310	36	512·04
0·8437	12	170·68	2·6014	37	526·26
0·9140	13	184·90	2·6717	38	540·49
0·9843	14	199·13	2·7420	39	554·71
1·0546	15	213·35	2·8123	40	568·93
1·1249	16	227·57	2·8826	41	583·16
1·1952	17	241·80	2·9529	42	597·38
1·2655	18	256·02	3·0232	43	611·60
1·3358	19	270·24	3·0935	44	625·83
1·4061	20	284·47	3·1638	45	640·05
1·4764	21	298·69	3·2341	46	654·27
1·5467	22	312·91	3·3044	47	668·50
1·6171	23	327·14	3·3747	48	682·72
1·6874	24	341·36	3·4450	49	696·94
1·7577	25	355·58	3·5153	50	711·17

KG./SQ. CM. — LB./SQ. IN.

kg./sq. cm.		lb./sq. in.	kg./sq. cm.		lb./sq. in.
3·5857	51	725·39	5·3433	76	1080·97
3·6560	52	739·61	5·4136	77	1095·20
3·7263	53	753·84	5·4839	78	1109·42
3·7966	54	768·06	5·5543	79	1123·64
3·8669	55	782·28	5·6246	80	1137·87
3·9372	56	796·51	5·6949	81	1152·09
4·0075	57	810·73	5·7652	82	1166·31
4·0778	58	824·95	5·8355	83	1180·54
4·1481	59	839·18	5·9058	84	1194·76
4·2184	60	853·40	5·9761	85	1208·98
4·2887	61	867·62	6·0464	86	1223·21
4·3590	62	881·85	6·1167	87	1237·43
4·4293	63	896·07	6·1870	88	1251·65
4·4996	64	910·29	6·2573	89	1265·88
4·5699	65	924·52	6·3276	90	1280·10
4·6403	66	938·74	6·3979	91	1294·32
4·7106	67	952·96	6·4682	92	1308·55
4·7809	68	967·19	6·5385	93	1322·77
4·8512	69	981·41	6·6089	94	1336·99
4·9215	70	995·63	6·6792	95	1351·22
4·9918	71	1009·86	6·7495	96	1365·44
5·0621	72	1024·08	6·8198	97	1379·66
5·1324	73	1038·30	6·8901	98	1393·89
5·2027	74	1052·53	6·9604	99	1408·11
5·2330	75	1066·75	7·0307	100	1422·33

TABLE OF STRESS VALUES

American	Metric	British
Pounds per square inch	Kilograms per square millimetre	British tons per square inch
250	0·175	0·112
500	0·350	0·222
1000	0·700	0·446
2000	1·400	0·893
3000	2·101	1·339
4000	2·801	1·786
5000	3·515	2·232
10000	7·031	4·464
15000	10·546	6·696
20000	14·062	8·929
25000	17·577	11·161
30000	21·092	13·393
35000	24·608	15·652
40000	28·123	17·875
45000	31·639	20·089
50000	35·154	22·321
55000	38·670	24·554
60000	42·185	26·786
65000	45·700	29·0!8
70000	49·216	31·250
75000	52·731	33·482
80000	56·247	35·594
85000	59·762	37·946
90000	63·277	40·179
95000	66·793	42·411
100000	70·308	44·643

TEMPERATURE CONVERSIONS

Albert Sauveur type of table. Look up reading in middle column; if in degrees Centigrade, read Fahrenheit equivalent in right-hand column; if in degrees Fahrenheit, read Centigrade equivalent in left-hand column. Values as printed in "Bethlehem Alloy Steels."

—459·4 to 0			0 to 100					
C.		F.	C.		F.	C.		F.
—273	—459·4		—17·8	0	32	10·0	50	122·0
—268	—450		—17·2	1	33·8	10·6	51	123·8
—262	—440		—16·7	2	35·6	11·1	52	125·6
—257	—430		—16·1	3	37·4	11·7	53	127·4
—251	—420		—15·6	4	39·2	12·2	54	129·2
—246	—410		—15·0	5	41·0	12·8	55	131·0
—240	—400		—14·4	6	42·8	13·3	56	132·8
—234	—390		—13·9	7	44·6	13·9	57	134·6
—229	—380		—13·3	8	46·4	14·4	58	136·4
—223	—370		—12·8	9	48·2	15·0	59	138·2
—218	—360		—12·2	10	50·0	15·6	60	140·0
—212	—350		—11·7	11	51·8	16·1	61	141·8
—207	—340		—11·1	12	53·6	16·7	62	143·6
—201	—330		—10·6	13	55·4	17·2	63	145·4
—196	—320		—10·0	14	57·2	17·8	64	147·2
—190	—310		— 9·4	15	59·0	18·3	65	149·0
—184	—300		— 8·9	16	60·8	18·9	66	150·8
—179	—290		— 8·3	17	62·6	19·4	67	152·6
—173	—280		— 7·8	18	64·4	20·0	68	154·4
—169	—273	—459	— 7·2	19	66·2	20·6	69	156·2
—168	—270	—454	— 6·7	20	68·0	21·1	70	158·0
—162	—260	—436	— 6·1	21	69·8	21·7	71	159·8
—157	—250	—418	— 5·6	22	71·6	22·2	72	161·6
—151	—240	—400	— 5·0	23	73·4	22·8	73	163·4
—146	—230	—382	— 4·4	24	75·2	23·3	74	165·2
—140	—220	—364	— 3·9	25	77·0	23·9	75	167·0
—134	—210	—346	— 3·3	26	78·8	24·4	76	168·8
—129	—200	—328	— 2·8	27	80·6	25·0	77	170·6
—123	—190	—310	— 2·2	28	82·4	25·6	78	172·4
—118	—180	—292	— 1·7	29	84·2	26·1	79	174·2
—112	—170	—274	— 1·1	30	86·0	26·7	80	176·0
—107	—160	—256	— 0·6	31	87·8	27·2	81	177·8
—101	—150	—238	0·0	32	89·6	27·8	82	179·6
— 96	—140	—220	0·6	33	91·4	28·3	83	181·4
— 90	—130	—202	1·1	34	93·2	28·9	84	183·2
— 84	—120	—184	1·7	35	95·0	29·4	85	185·0
— 79	—110	—166	2·2	36	96·8	30·0	86	186·8
— 73	—100	—148	2·8	37	98·6	30·6	87	188·6
— 68	— 90	—130	3·3	38	100·4	31·1	88	190·4
— 62	— 80	—112	3·9	39	102·2	31·7	89	192·2
— 57	— 70	— 94	4·4	40	104·0	32·2	90	194·0
— 51	— 60	— 76	5·0	41	105·8	32·8	91	195·8
— 46	— 50	— 58	5·6	42	107·6	33·3	92	197·6
— 40	— 40	— 40	6·1	43	109·4	33·9	93	199·4
— 34	— 30	— 22	6·7	44	111·2	34·4	94	201·2
— 29	— 20	— 4	7·2	45	113·0	35·0	95	203·0
— 23	— 10	14	7·8	46	114·8	35·6	96	204·8
— 17	0	32	8·3	47	116·6	36·1	97	206·6
			8·9	48	118·4	36·7	98	208·4
			9·4	49	120·2	37·2	99	210·2
						37·8	100	212·0

100 to 1000						1000 to 2000					
C.		F.	C.		F.	C.		F.	C.		F.
38	100	212	260	500	932	538	1000	1832	816	1500	2732
43	110	230	266	510	950	543	1010	1850	821	1510	2750
49	120	248	271	520	968	549	1020	1868	827	1520	2768
54	130	266	277	530	986	554	1030	1886	832	1530	2786
60	140	284	282	540	1004	560	1040	1904	838	1540	2804
66	150	302	288	550	1022	566	1050	1922	843	1550	2822
71	160	320	293	560	1040	571	1060	1940	849	1560	2840
77	170	338	299	570	1058	577	1070	1958	854	1570	2858
82	180	356	304	580	1076	582	1080	1976	860	1580	2876
88	190	374	310	590	1094	588	1090	1994	866	1590	2894
93	200	392	316	600	1112	593	1100	2012	871	1600	2912
99	210	410	321	610	1130	599	1110	2030	877	1610	2930
100	212	414	327	620	1148	604	1120	2048	882	1620	2948
104	220	428	332	630	1166	610	1130	2066	888	1630	2966
110	230	446	338	640	1184	616	1140	2084	893	1640	2984
116	240	464	343	650	1202	621	1150	2102	899	1650	3002
121	250	482	349	660	1220	627	1160	2120	904	1660	3020
127	260	500	354	670	1238	632	1170	2138	910	1670	3038
132	270	518	360	680	1256	638	1180	2156	916	1680	3056
138	280	536	366	690	1274	643	1190	2174	921	1690	3074
143	290	554	371	700	1292	649	1200	2192	927	1700	3092
149	300	572	377	710	1310	654	1210	2210	932	1710	3110
154	310	590	382	720	1328	660	1220	2228	938	1720	3128
160	320	608	388	730	1346	666	1230	2246	943	1730	3146
166	330	626	393	740	1364	671	1240	2264	949	1740	3164
171	340	644	399	750	1382	677	1250	2282	954	1750	3182
177	350	662	404	760	1400	682	1260	2300	960	1760	3200
182	360	680	410	770	1418	688	1270	2318	966	1770	3218
188	370	698	416	780	1436	693	1280	2336	971	1780	3236
193	380	716	421	790	1454	699	1290	2354	977	1790	3254
199	390	734	427	800	1472	704	1300	2372	982	1800	3272
204	400	752	432	810	1490	710	1310	2390	988	1810	3290
210	410	770	438	820	1508	716	1320	2408	993	1820	3308
216	420	788	443	830	1526	721	1330	2426	999	1830	3326
221	430	806	449	840	1544	727	1340	2444	1004	1840	3344
227	440	824	454	850	1562	732	1350	2462	1010	1850	3362
232	450	842	460	860	1580	738	1360	2480	1016	1860	3380
238	460	860	466	870	1598	743	1370	2498	1021	1870	3398
243	470	878	471	880	1616	749	1380	2516	1027	1880	3416
249	480	896	477	890	1634	754	1390	2534	1032	1890	3434
254	490	914	482	900	1652	760	1400	2552	1038	1900	3452
			488	910	1670	766	1410	2570	1043	1910	3470
			493	920	1688	771	1420	2588	1049	1920	3488
			499	930	1706	777	1430	2606	1054	1930	3506
			504	940	1724	782	1440	2624	1060	1940	3524
			510	950	1742	788	1450	2642	1066	1950	3542
			516	960	1760	793	1460	2660	1071	1960	3560
			521	970	1778	799	1470	2678	1077	1970	3578
			527	980	1796	804	1480	2696	1082	1980	3596
			532	990	1814	810	1490	2714	1088	1990	3614
			538	1000	1832				1093	2000	3632

2000 to 3000									
C.	F.	C.	F.	C.	F.	C.	F.	C.	F.
1093	2000	3632	1232	2250	4082	1371	2500	4532	1510	2750	4982
1099	2010	3650	1238	2260	4100	1377	2510	4550	1516	2760	5000
1104	2020	3668	1243	2270	4118	1382	2520	4568	1521	2770	5018
1110	2030	3686	1249	2280	4136	1388	2530	4586	1527	2780	5036
1116	2040	3704	1254	2290	4154	1393	2540	4604	1532	2790	5054
1121	2050	3722	1260	2300	4172	1399	2550	4622	1538	2800	5072
1127	2060	3740	1266	2310	4190	1404	2560	4640	1543	2810	5090
1132	2070	3758	1271	2320	4208	1410	2570	4658	1549	2820	5108
1138	2080	3776	1277	2330	4226	1416	2580	4676	1554	2830	5126
1143	2090	3794	1282	2340	4244	1421	2590	4694	1560	2840	5144
1149	2100	3812	1288	2350	4262	1427	2600	4712	1566	2850	5162
1154	2110	3830	1293	2360	4280	1432	2610	4730	1571	2860	5180
1160	2120	3848	1299	2370	4298	1438	2620	4748	1577	2870	5198
1166	2130	3866	1304	2380	4316	1443	2630	4766	1582	2880	5216
1171	2140	3884	1310	2390	4334	1449	2640	4784	1588	2890	5234
1177	2150	3902	1316	2400	4352	1454	2650	4802	1593	2900	5252
1182	2160	3920	1321	2410	4370	1460	2660	4820	1599	2910	5270
1188	2170	3938	1327	2420	4388	1466	2670	4838	1604	2920	5288
1193	2180	3956	1332	2430	4406	1471	2680	4856	1610	2930	5306
1199	2190	3974	1338	2440	4424	1477	2690	4874	1616	2940	5324
1204	2200	3992	1343	2450	4442	1482	2700	4892	1621	2950	5342
1210	2210	4010	1349	2460	4460	1488	2710	4910	1627	2960	5360
1216	2220	4028	1354	2470	4478	1493	2720	4928	1632	2970	5378
1221	2230	4046	1360	2480	4496	1499	2730	4946	1638	2980	5396
1227	2240	4064	1366	2490	4514	1504	2740	4964	1643	2990	5414
									1649	3000	5432

HYDROMETER CONVERSION TABLES

Showing the Relation between Density (C.G.S.) and the Baumé and Twaddell Scales for Densities above Unity

Density	Degrees Baumé	Degrees Twaddell	Density	Degrees Baumé	Degrees Twaddell
1·00	0·00	0	1·41	42·16	82
1·01	1·44	2	1·42	42·89	84
1·02	2·84	4	1·43	43·60	86
1·03	4·22	6	1·44	44·31	88
1·04	5·58	8	1·45	45·00	90
1·05	6·91	10	1·46	45·68	92
1·06	8·21	12	1·47	46·36	94
1·07	9·49	14	1·48	47·03	96
1·08	10·74	16	1·49	47·68	98
1·09	11·97	18	1·50	48·33	100
1·10	13·18	20	1·51	48·97	102
1·11	14·37	22	1·52	49·60	104
1·12	15·54	24	1·53	50·23	106
1·13	16·68	26	1·54	50·84	108
1·14	17·81	28	1·55	51·45	110
1·15	18·91	30	1·56	52·05	112
1·16	20·00	32	1·57	52·64	114
1·17	21·07	34	1·58	53·23	116
1·18	22·12	36	1·59	53·80	118
1·19	23·15	38	1·60	54·38	120
1·20	24·17	40	1·61	54·94	122
1·21	25·16	42	1·62	55·49	124
1·22	26·15	44	1·63	56·04	126
1·23	27·11	46	1·64	56·58	128
1·24	28·06	48	1·65	57·12	130
1·25	29·00	50	1·66	57·65	132
1·26	29·92	52	1·67	58·17	134
1·27	30·83	54	1·68	58·69	136
1·28	31·72	56	1·69	59·20	138
1·29	32·60	58	1·70	59·71	140
1.30	33·46	60	1·71	60·20	142
1·31	34·31	62	1·72	60·70	144
1·32	35·15	64	1·73	61·18	146
1·33	35·98	66	1·74	61·67	148
1·34	36·79	68	1·75	62·14	150
1·35	37·59	70	1·76	62·61	152
1·36	38·38	72	1·77	63·08	154
1·37	39·16	74	1·78	63·54	156
1·38	39·93	76	1·79	63·99	158
1·39	40·68	78	1·80	64·44	160
1·40	41·43	80	—	...	...

METRIC AND DECIMAL EQUIVALENTS OF FRACTIONS

(OF ONE INCH)

mm.							inches
0·3969						$\frac{1}{64}$	0·015625
0·79375					$\frac{1}{32}$		0·03125
1·1906						$\frac{3}{64}$	0·046875
1·5875				$\frac{1}{16}$			0·0625
1·9844						$\frac{5}{64}$	0·078125
2·38125					$\frac{3}{32}$		0·09375
2·7781						$\frac{7}{64}$	0·109375
3·1750			$\frac{1}{8}$				0·125
3·5719						$\frac{9}{64}$	0·140625
3·96875					$\frac{5}{32}$		0·15625
4·3656						$\frac{11}{64}$	0·171875
4·7625				$\frac{3}{16}$			0·1875
5·1594						$\frac{13}{64}$	0·203125
5·55625					$\frac{7}{32}$		0·21875
5·9531						$\frac{15}{64}$	0·234375
6·3500		$\frac{1}{4}$					0·25
6·7469						$\frac{17}{64}$	0·265625
7·14375					$\frac{9}{32}$		0·28125
7·5406						$\frac{19}{64}$	0·296875
7·9375				$\frac{5}{16}$			0·3125
8·3344						$\frac{21}{64}$	0·328125
8·73125					$\frac{11}{32}$		0·34375
9·1281						$\frac{23}{64}$	0·359375
9·5250			$\frac{3}{8}$				0·375
9·9219						$\frac{25}{64}$	0·390625
10·31875					$\frac{13}{32}$		0·40625
10·7156						$\frac{27}{64}$	0·421875
11·1125				$\frac{7}{16}$			0·4375
11·5094						$\frac{29}{64}$	0·453125
11·90625					$\frac{15}{32}$		0·46875
12·3031						$\frac{31}{64}$	0·484375
12·7000	$\frac{1}{2}$						0·5

METRIC AND DECIMAL EQUIVALENTS OF FRACTIONS

(OF ONE INCH)

mm.							inches
13·0969						$\frac{33}{64}$	0·515625
13·49375					$\frac{17}{32}$		0·53125
13·8906						$\frac{35}{64}$	0·546875
14·2875				$\frac{9}{16}$			0·5625
14·6844						$\frac{37}{64}$	0·578125
15·08125					$\frac{19}{32}$		0·59375
15·4781						$\frac{39}{64}$	0·609375
15·8750			$\frac{5}{8}$				0·625
16·2719						$\frac{41}{64}$	0·640625
16·66875					$\frac{21}{32}$		0·65625
17·0656						$\frac{43}{64}$	0·671875
17·4625				$\frac{11}{16}$			0·6875
17·8594						$\frac{45}{64}$	0·703125
18·25625					$\frac{23}{32}$		0·71875
18·6531						$\frac{47}{64}$	0·734375
19·0500		$\frac{3}{4}$					0·75
19·4469						$\frac{49}{64}$	0·765625
19·84375					$\frac{25}{32}$		0·78125
20·2406						$\frac{51}{64}$	0·796875
20·6375				$\frac{13}{16}$			0·8125
21·0344						$\frac{53}{64}$	0·828125
21·43125					$\frac{27}{32}$		0·84375
21·8281						$\frac{55}{64}$	0·859375
22·2250			$\frac{7}{8}$				0·875
22·6219						$\frac{57}{64}$	0·890625
23·01875					$\frac{29}{32}$		0·90625
23·4156						$\frac{59}{64}$	0·921875
23·8125				$\frac{15}{16}$			0·9375
24·2094						$\frac{61}{64}$	0·953125
24·60625					$\frac{31}{32}$		0·96875
25·0031						$\frac{63}{64}$	0·984375
25·4000	1						1·0

AREAS AND CIRCUMFERENCES OF CIRCLES

(See note at end of tables)

Diameter	Circumference	Area	Diameter	Circumference	Area
$\frac{1}{16}$	0·1964	0·0031	$\frac{3}{4}$	5·4978	2·4053
0·1	0·3143	0·0078	1·8	5·6548	2·5446
$\frac{1}{8}$	0·3929	0·0123	$\frac{7}{8}$	5·8905	2·7612
$\frac{3}{16}$	0·5893	0·0276	·9	5·9690	2·8352
0·2	0·6286	0·0314			
$\frac{1}{4}$	0·7857	0·0491	2·0	6·2832	3·1416
0·3	0·9429	0·0707	·1	6·5973	3·4636
$\frac{5}{16}$	0·9821	0·0767	$\frac{1}{8}$	6·6759	3·5456
$\frac{3}{8}$	1·1786	0·1105	·2	6·9115	3·8013
0·4	1·2571	0·1257	$\frac{1}{4}$	7·0686	3·9761
$\frac{7}{16}$	1·3750	0·1504	·3	7·2256	4·1547
$\frac{1}{2}$	1·5714	0·1964	$\frac{3}{8}$	7·4613	4·4301
$\frac{9}{16}$	1·7679	0·2486	·4	7·5398	4·5239
0·6	1·8857	0·2829	$\frac{1}{2}$	7·8540	4·9087
$\frac{5}{8}$	1·9643	0·3069	·6	8·1681	5·3093
$\frac{11}{16}$	2·1607	0·3714	$\frac{5}{8}$	8·2467	5·4119
0·7	2·2000	0·3850	·7	8·4823	5·7255
$\frac{3}{4}$	2·3571	0·4419	$\frac{3}{4}$	8·6394	5·9396
0·8	2·5143	0·5029	·8	8·7964	6·1575
$\frac{13}{16}$	2·5536	0·5187	$\frac{7}{8}$	9·0321	6·4918
$\frac{7}{8}$	2·7500	0·6016	·9	9·1106	6·6052
0·9	2·8286	0·6364			
$\frac{15}{16}$	2·9464	0·6906	3·0	9·4248	7·0686
			·1	9·7389	7·5476
1·0	3·1416	0·7854	$\frac{1}{8}$	9·8175	7·6699
·1	3·4557	0·9503	·2	10·1531	8·0424
$\frac{1}{8}$	3·5343	0·9940	$\frac{1}{4}$	10·2102	8·2958
·2	3·7699	1·1309	·3	10·3672	8·5530
$\frac{1}{4}$	3·9270	1·2272	$\frac{3}{8}$	10·6029	8·9462
·3	4·0840	1·3273	·4	10·6814	9·0792
$\frac{3}{8}$	4·3197	1·4849	$\frac{1}{2}$	10·9956	9·6211
·4	4·3982	1·5393	·6	11·3197	10·178
$\frac{1}{2}$	4·7124	1·7671	$\frac{5}{8}$	11·3883	10·321
·6	5·0265	2·0106	·7	11·6239	10·752
$\frac{5}{8}$	5·1051	2·0739	$\frac{3}{4}$	11·7810	11·045
·7	5·3407	2·2698	·8	11·9380	11·341

(See note at end of tables)

Diameter	Circumference	Area	Diameter	Circumference	Area
$\frac{7}{8}$	12·1737	11·793	6·0	18·8496	28·274
3·9	12·2522	11·945	·1	19·1637	29·224
			$\frac{1}{8}$	19·2423	29·464
4·0	12·5664	12·566	·2	19·4779	30·190
·1	12·8805	13·202	$\frac{1}{4}$	19·6350	30·679
$\frac{1}{8}$	12·9591	13·364	·3	19·7930	31·272
·2	13·2947	13·854	$\frac{3}{8}$	20·0277	31·919
$\frac{1}{4}$	13·3518	14·186	·4	20·1062	32·169
·3	13·5088	14·522	$\frac{1}{2}$	20·4204	33·183
$\frac{3}{8}$	13·7445	15·033	·6	20·7345	34·212
·4	13·8230	15·205	$\frac{5}{8}$	20·8131	34·471
$\frac{1}{2}$	14·1372	15·904	·7	21·0487	35·256
·6	14·4513	16·619	$\frac{3}{4}$	21·2058	35·784
$\frac{5}{8}$	14·5299	16·800	·8	21·4628	36·316
·7	14·7555	17·349	$\frac{7}{8}$	21·5985	37·122
$\frac{3}{4}$	14·9226	17·720	·9	21·6770	37·392
·8	15·0796	18·095			
$\frac{7}{8}$	15·3153	18·665	7·0	21·991	38·484
·9	15·3938	18·857	·1	22·305	39·592
			$\frac{1}{8}$	22·383	39·871
5·0	15·7080	19·635	·2	22·619	40·715
·1	16·0221	20·428	$\frac{1}{4}$	22·776	41·282
$\frac{1}{8}$	16·1007	20·629	·3	22·933	41·853
·2	16·3363	21·237	$\frac{3}{8}$	23·169	42·718
$\frac{1}{4}$	16·4934	21·647	·4	23·247	43·008
·3	16·6504	22·061	$\frac{1}{2}$	23·562	44·178
$\frac{3}{8}$	16·8861	22·690	·6	23·876	45·364
·4	16·9646	22·902	$\frac{5}{8}$	23·954	45·663
$\frac{1}{2}$	17·2788	23·758	·7	24·190	46·566
·6	17·5929	24·630	$\frac{3}{4}$	24·347	47·173
$\frac{5}{8}$	17·6715	24·850	·8	24·504	47·783
·7	17·9071	25·517	$\frac{7}{8}$	24·740	48·707
$\frac{3}{4}$	18·0642	25·967	·9	24·818	49·016
·8	18·2212	26·420			
$\frac{7}{8}$	18·4569	27·108	8·0	25·132	50·265
·9	18·5354	27·339	·1	25·446	51·530

Area and Circumferences of Circles—continued

(See note at end of tables)

Diameter	Circumference	Area	Diameter	Circumference	Area
$\frac{1}{8}$	25·525	51·848	$\frac{1}{4}$	32·201	82·516
8·2	25·761	52·810	10·3	32·358	83·323
$\frac{1}{4}$	25·918	53·456	$\frac{3}{8}$	32·594	84·540
·3	26·075	54·106	·4	32·672	84·948
$\frac{3}{8}$	26·310	55·088	$\frac{1}{2}$	32·986	86·590
·4	26·389	55·417	·6	33·300	88·247
$\frac{1}{2}$	26·703	56·745	$\frac{5}{8}$	33·379	88·664
·6	27·017	58·088	·7	33·615	89·920
$\frac{5}{8}$	27·096	58·426	$\frac{3}{4}$	33·772	90·762
·7	27·331	59·446	·8	33·929	91·609
$\frac{3}{4}$	27·489	60·132	$\frac{7}{8}$	34·164	92·885
·8	27·646	60·821	·9	34·243	93·313
$\frac{7}{8}$	27·881	61·862			
·9	27·960	62·211	$11\frac{1}{4}$	35·343	99·40
			$\frac{1}{2}$	36·128	103·87
9·0	28·274	63·617	$\frac{3}{4}$	36·913	108·44
·1	28·588	65·038	$12\frac{1}{4}$	38·484	117·86
$\frac{1}{8}$	28·667	65·396	$\frac{1}{2}$	39·270	122·72
·2	28·902	66·476	$\frac{3}{4}$	40·055	127·68
$\frac{1}{4}$	29·059	67·200	$13\frac{1}{4}$	41·626	137·89
·3	29·216	67·929	$\frac{1}{2}$	42·411	143·14
$\frac{3}{8}$	29·452	69·029	$\frac{3}{4}$	43·197	148·49
·4	29·531	69·397	$14\frac{1}{4}$	44·767	159·49
$\frac{1}{2}$	29·845	70·882	$\frac{1}{2}$	45·553	165·13
·6	30·159	72·382	$\frac{3}{4}$	46·338	170·87
$\frac{5}{8}$	30·237	72·759	$15\frac{1}{4}$	47·909	182·65
·7	30·473	73·398	$\frac{1}{2}$	48·694	188·69
$\frac{3}{4}$	30·630	74·662	$\frac{3}{4}$	49·480	194·83
·8	30·787	75·429	$16\frac{1}{4}$	51·051	207·39
$\frac{7}{8}$	31·023	76·588			
·9	31·101	76·977	$\frac{1}{2}$	51·836	213·82
			$\frac{3}{4}$	52·621	220·35
10·0	31·416	78·540	$17\frac{1}{4}$	54·192	233·71
·1	31·730	80·118	$\frac{1}{2}$	54·978	240·52
$\frac{1}{8}$	31·808	80·515	$\frac{3}{4}$	55·763	247·45
·2	32·044	81·713	$18\frac{1}{4}$	57·334	261·59

Diameter	Circum-ference	Area	Diameter	Circum-ference	Area
1/8	58·119	268·80	20¼	63·617	322·06
¼	58·905	276·12	½	64·402	330·06
19⅜	60·475	291·04	¾	65·188	338·16
½	61·261	298·65	21·0	65·973	346·36
¾	62·046	306·36	22·0	69·115	380·13

NOTES

From these tables circumferences and areas of circles with a diameter of any whole number (ins., mm., etc.) up to 110 can be obtained. Thus, the circumference of 74 is read from the value opposite 7·4 and is 232·47, the decimal point being moved one place to the right. The corresponding area is 4,300·8, the point being moved two places to the right.

Diameters are given in 8ths and 4ths, as well as 10ths, so that British units up to 21 can be dealt with. The whole numbers not given, e.g., 11, 12, etc., are read from the first part of the table, since it will be clear that the figures, for instance, opposite the last diameters, 21 and 22, are identical with those opposite 2·1 and 2·2, except for the positions of the decimal points.

Approximate Weights of Materials in Pounds Per Cubic Foot

Aluminium Cast ...	160
Aluminium Wrought ...	167
Aluminium Bronze ...	475
Ashes	37
Brass Rolled	524
Brass Swarf	157
Babbit Metal	454
Brick, Common ...	85–118
Bronze	534
Cast Iron (mean) ...	450
Cast Iron Turnings ...	140
Cement (loose)	85
Chalk	140–145
Charcoal (lump) ...	18
Clay	120–135
Coal	60–80
Coal Dust	53
Coke	28
Concrete (set cement) ...	140
Copper Cast	548
Cupola Slag	150
Fire Clay	90
Fire brick (in mortar)	112–125
French Chalk	162
Gold (cast pure)	1,200
Gold 22 carat	1,090

Lead	710
Limestone ...	158–168
Mercury	847
Monel	554
Nickel Cast	516
Nickel Silver	516
Oak Wood	53·5
Phosphor Bronze Cast ...	536
Pine Wood	30
Pig Iron (mean)	300
Pig Iron and Scrap as charged into the cupola	336
Plumbago (graphite)	137–145
Sand (moulding), fairly dry and unrammed ...	75–85
Sand (silica)	85–90
Silver Pure Cast ...	656
Steel (mean)	490
Tin Pure	453
Water—as ice	58·7
Water at 32° F. ...	62·4
Water at 212° F. ...	59·6
Wrought Iron	480
Zinc Cast...	428
Zinc Rolled	448

SOLDERS AND FUSIBLE ALLOYS

Fusible Alloys

Trade Name or Type of Alloy	Composition (%)					Solidus		Liquidus	
	Sn	Bi	Pb	Cd	Others	°C	°F	°C	°F
Binary Eutectic	99·25	—	—	—	0·75 Cu	227	441	227	441
Binary Eutectic	96·5	—	—	—	3·5 Ag	221	430	221	430
Binary Eutectic	—	—	—	17·0	83·0 Tl	203	397	203	397
Binary Eutectic	92·0	—	—	—	8·0 Zn	199	390	199	390
Binary Eutectic	—	47·5	—	—	52·5 Tl	188	370	188	370
Binary Eutectic	62·0	—	38·0	—	—	183	361	183	361
Binary Eutectic	67·0	—	—	33·0	—	176	349	176	349
Binary Eutectic	56·5	—	—	—	43·5 Tl	170	338	170	338
Ternary Eutectic	40·0	—	42·0	18·0	—	145	293	160	320
Ternary Eutectic	51·2	—	30·6	18·2	—	145	293	145	293
Binary Eutectic	—	60·0	—	40·0	—	144	291	144	291
Ternary Eutectic	48·8	10·2	41·0	—	—	142	288	166	331
Binary Eutectic—Cerrotru	43·0	57·0	—	—	—	138	281	138	281
Ternary Eutectic	41·6	57·4	1·0	—	—	134	273	135	275
Ternary Eutectic	40·0	56·0	—	—	4·0 Zn	130	266	130	266
Ternary Eutectic	46·0	—	—	17·0	37·0 Tl	128	262	128	262
Binary Eutectic—Cerrobase	—	55·5	44·5	—	—	124	255	124	255
Binary Eutectic	48·0	—	—	—	52·0 In	117	243	117	243
Cerroseal—35	50·0	—	—	—	50·0 In	117	243	127	260
	1·0	55·0	44·0	—	—	117	243	120	248

	Sn	Bi	Pb	Cd	Other	°C	°F	°C	°F
Cerromatrix	14·5	48·0	28·5	—	9·0 Sb	103	217	227	440
	34·5	44·5	—	21·0	—	103	217	120	248
	25·0	50·0	—	25·0	—	103	217	113	235
Ternary Eutectic	25·9	53·9	—	20·2	—	103	217	103	217
	33·0	34·0	33·0	—	—	96	205	143	289
Malotte's	34·2	46·1	19·7	—	—	96	205	123	253
Rose's	22·0	50·0	28·0	—	—	96	205	110	230
D'Arcet's	25·0	50·0	25·0	—	—	96	205	98	208
Newton's	18·8	50·0	31·2	—	—	96	205	97	207
Onion's or Lichtenberg's	20·0	50·0	30·0	—	—	96	205	100	212
Ternary Eutectic	15·5	52·5	32·0	—	—	92	198	96	205
Ternary Eutectic	—	51·7	40·2	8·1	—	92	198	92	198
	15·4	38·4	30·8	15·4	—	70	158	97	207
Cerrosafe	11·3	42·5	37·7	8·5	—	70	158	90	194
	24·5	45·3	17·9	12·3	—	70	158	88	190
	13·0	40·0	37·0	10·0	—	70	158	85	185
	13·0	42·0	35·0	10·0	—	70	158	80	176
Lipowitz's or Cerrobend	13·3	50·0	26·7	10·0	—	70	158	73	163
Wood's	12·5	50·0	25·0	12·5	—	70	158	72	162
Quaternary Eutectic	13·1	49·5	27·3	10·1	—	70	158	70	158
	13·2	49·3	26·3	9·8	1·4 Ga	65	149	66	151
Cerrolow 147	12·77	48·0	25·6	9·6	4·0 In	61	142	65	149
Cerrolow 136B	15·0	49·0	18·0		18·0 In	57	136	69	156
Cerrolow 136	12·0	49·0	18·0		21·0 In	57	136	57	136
Cerrolow 140	12·6	47·5	25·4	9·5	19·1 In	56	134	65	149
Cerrolow 117	8·3	44·7	22·6	5·3	16·1 In	47	117	47	117
Cerrolow 117B	11·3	44·7	22·6	5·3	18·33 In:4 Hg	47	117	52	126
Cerrolow 105	7·97	42·91	21·7	5·09		38	100	43	110
Binary Eutectic	8·0	—	—	—	92·0 Ga	20	68	20	68

(Courtesy Tin Research Institute)

WEIGHT OF CASTING FROM WEIGHT OF PATTERN

Allowance must be made for the weight of any metal in the pattern.
The patterns are without cores.

A pattern weighing 1-lb. when made of	Cast Iron	Cast Steel	Yellow Brass	Gunmetal or Bronze	Bell Bronze	Zinc	Copper	Aluminium
Alder ...	12·8	14·6	14·3	14·8	15·5	12·2	15·6	4·6
Baywood ...	8·8	9·8	9·9	10·3	14·0	8·5	10·5	3·2
Beechwood	8·5	9·5	9·5	10·0	12·0	8·2	10·1	3·1
Birch ...	10·6	11·2	11·9	12·3	12·9	10·2	12·0	3·9
Cedar ...	12·5	14·5	14·2	14·7	15·3	12·0	15·7	4·5
Cherry ...	10·7	12·0	12·0	12·6	13·5	10·4	12·8	3·9
Lime ...	13·4	14·2	15·1	15·6	16·3	12·9	15·3	4·9
Linden ...	12·0	13·3	13·5	14·1	15·0	11·6	14·3	4·3
Mahogany ...	8·5	9·5	9·5	10·0	10·5	8·2	10·1	3·1
Maple ...	9·2	10·3	10·3	10·6	10·9	8·9	11·0	3·2
Oak ...	9·4	10·4	10·5	10·8	11·0	9·1	11·2	3·4
Pear ...	10·9	12·1	12·2	12·8	14·0	10·6	13·0	3·9
Pine ...	14·7	16·3	16·5	16·6	17·3	14·3	17·5	5·3
Whitewood ...	16·4	18·1	18·4	19·3	20·0	15·9	19·5	5·9
Brass ...	0·84	0·98	0·95	0·99	1·0	0·81	1·04	0·31
Iron ...	0·97	1·09	1·09	1·13	1·18	0·93	1·17	0·35
Lead ...	0·64	0·75	0·72	0·74	0·78	0·61	0·8	0·23
Tin (⅓-¼ lead) ...	0·89	1·1	1·00	1·00	1·12	0·85	1·18	0·32
Zinc ...	1·00	1·16	1·13	1·17	1·22	0·96	1·25	0·36

Will weigh when cast in

SHRINKAGE AND CONTRACTION OF CASTING ALLOYS

Volume Changes on Cooling. Although shrinkage and contraction are more or less synonymous terms they have special application with respect to castings. Three stages of volume changes are recognised.

A. Liquid shrinkage which takes place during the fall in temperature from pouring to freezing.

B. Shrinkage during freezing or solidification which with some alloys may be offset to some extent by an accompanying expansion.

C. Solid shrinkage or contraction which occurs in the solidified casting as it cools to atmospheric temperature.

The following table gives some idea of the reduction in volume which takes place in various casting alloys during cooling from an average pouring temperature to their freezing point.

	Liquid Shrinkage Volume per cent.
Soft grey iron (3·5% to 4·0% total carbon)	0·6
Aluminium alloy (11% to 13% silicon)	3·5
Copper (deoxidised)	3·8
Aluminium bronze (90% copper, 10% aluminium) ...	4·1
Grey iron (3·0% to 3·25% total carbon)	4·2
Manganese bronze (40% zinc, 1·25% iron, 1% aluminium, 0·5% manganese, 0·5% tin)	4·6
Grey iron, high duty (2·5% to 2·75% total carbon)	5·0
Nickel silver (20% nickel, 15% zinc, 65% copper) ...	5·5
Ni-resist austenitic cast iron	5·6
White iron, chill cast	5·75
Nickel (98% nickel, 1·5% silicon, 0·1% carbon) ...	6·1
Monel (1·0% silicon)	6·3
Yellow Brass (27% zinc, 2% lead, 1% tin)	6·4
Nickel silver (20% nickel, 15% zinc, 4% tin, 5% lead)	6·5
Aluminium rich alloys not containing silicon ...	6·5 to 8·0
Steel	7·22
White iron, malleable	7·25
Bearing bronze (10% tin, 10% lead)	7·3

These values are a summation of the total volume change which occurs under headings A and B. Where coring is considerable and with "hard" cores, the expansion of the cores, and to some extent moulds, will neutralise some of the shrinkage tendencies. In some cases liquid metal may extrude through the risers or feeders as a result of such expansion.

To reduce shrinkage defects and porosity in castings the liquid and solidification shrinkage (A and B) must be made good by a supply of liquid metal from an adequate system of feeding. Some information on the design of feeding heads is given under a separate heading.

The shrinkage or contraction under heading C is the factor with which the pattern maker is concerned. Pattern dimensions are made to "contraction rule" and not "standard rule." Different casting alloys have different contraction rates. Although average values for unrestricted cooling conditions are available as set out in the following table taken from the *Metal Industry Handbook*, it must be borne in mind that these can vary considerably according to the resistance offered by moulds and cores.

Alloy		Patternmakers' Contraction Allowance
Aluminium alloys		
Commercially pure		
Al—Cu—Si alloys	e.g. LM4	
Al—Si alloys	e.g. LM6, LM28, LM29, LM30	
Al—Cu alloys	e.g. LM12	1·3% or 1/75
Al—Mg 5—10%	e.g. LM5, LM10	
Low expansion Al—Si—Ni	e.g. LM13	
Aluminium bronze		2·0—2·3%
Beryllium—Copper		1·6%
Bismuth		1·3%
Brass, yellow (thick)		1·3%
Brass, yellow (thin)		1·6%
*Cast iron, grey		0·9—1·3%
Cast iron, white		2·0%
Copper		1·6%
Delta Bronze		1·6%
Gunmetal		1·0—1·6%
Lead		2·6%
Magnesium		2·0%
Magnesium alloys (5%, or over of alloys, excluding cadmium, which has no effect on the shrinkage of magnesium)		1·6%
Manganese bronze		2·0%
Monel		2·0%
Nickel		2·0%
Nickel silver		1·0—1·6%
Phosphor—bronze		1·0—1·6%
Silicon bronze		1·3—1·6%
Steel, carbon		1·6—2·0%
Steel, chromium		2·0%
Steel, manganese		2·6%
Stone's gear wheel bronze		1·0%
Tin		2·0%
White metal		0·6%
Zinc and zinc-base alloys		2·6%

*The amount of contraction in cast iron depends chiefly upon the speed with which the casting cools, greater contraction resulting from more rapid cooling, and vice versa. Large hollow castings contract less in a vertical than in a horizontal direction.

73

AIR CONSUMPTION OF VARIOUS PNEUMATIC TOOLS

Taken from "Foundry"

Tool									
Small Spray Guns	2–3 cu. ft. per min.								
Jar Ram or Jolters (platform type)	30–40 cu. ft. per ton.								

Hand Grinders—

Wheel diam. ins.	$1\frac{1}{2}$	2	4	6	8
cu. ft. per min.	12	16	28–35	40–45	50–60

Hand Sanders

pad size, inches	7	9
cu. ft. per min.	35	45

Chipping Hammers   stroke, ins. $1\frac{1}{2}$ to 4 average 20–25 cu. ft per minute.

Scaling Hammers ...   average 15 cu. ft. per minute.

Air Hoists ...

capacity lb.	500–1000	2000	3000	4000	6000
cu. ft. per min.	0·5	2	3·2	6·3	8·4

Geared Hoists ...

capacity tons	1	$1\frac{1}{2}$	2	3	4	5	6	8	10
cu. ft. min. per ft. lift	3	5	6	8	10	15	20	25	30

Impact Wrenches

size ins.	$\frac{1}{4}$	$\frac{3}{8}$	$\frac{5}{8}$	$\frac{3}{4}$	$1\frac{1}{4}$	$1\frac{3}{4}$
cu. ft. per min.	8–9	12	22	28	38	50

Jacks ...

cyl. diam. ins.	8	10	12	14	16	20	24
cu. ft. min. per lift	1·8	2·8	4·0	5·0	6·9	11·1	16

Motors ...

horse power	2	4	5	8	15
cu. ft. per min.	40–50	60–70	90–100	140	240

Rotary Steel Drills

drill size ins.	$\frac{1}{4}$	$\frac{5}{8}$	$\frac{1}{2}$	$\frac{7}{8}$	1	$1\frac{1}{4}$–2
cu. ft. per min.	20	28–32	36–42	55	65	85–90

Wood Boring Drills

drill size ins.	$\frac{1}{2}$	$\frac{7}{8}$	1	$1\frac{1}{4}$
cu. ft. per min.	20	32	38	45

Note : Multiply cu. ft./min. by 472·0 to obtain cu. cm/sec.

HARDNESS CONVERSION CHART

Brinell — Rockwell — Scleroscope — Vickers

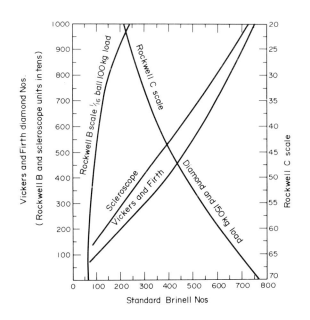

SEGER CONES

These are cones made of refractory materials which bend at a definite temperature on heating. The following table gives the bending temperature, in degrees centigrade, for each cone.

Approx. Temp. in degrees Centigrade	Cone No.	Approx. Temp. in degrees Centigrade	Cone No.	Approx. Temp. in degrees Centigrade	Cone No.
600	022	1080	01A	1560	26
650	021	1100	1A	1610	27
670	020	1120	2A	1630	28
690	019	1140	3A	1650	29
710	018	1160	4A	1670	30
730	017	1180	5A	1690	31
750	016	1200	6A	1710	32
790	015A	1230	7	1730	33
815	014A	1250	8	1750	34
835	013A	1280	9	1770	35
855	012A	1300	10	1790	36
880	011A	1320	11	1825	37
900	010A	1350	12	1850	38
920	09A	1380	13	1860	39
940	08A	1410	14	1920	40
960	07A	1435	15	1960	41
980	06A	1460	16	2000	42
1000	05A	1480	17		
1020	04A	1500	18		
1040	03A	1520	19		
1060	02A	1530	20		

(*Figures supplied by Wenger's Ltd., Etruria*)

USUAL THICKNESS OF CHILLS FOR CHILLED ROLLS

Diameter of Roll, Inches	Thickness of Chill, Inches	Diameter of Roll, Inches	Thickness of Chill, Inches	Diameter of Roll, Inches	Thickness of Chill, Inches
3	2½	13	5	23	8½
4	2½	14	5¼	24	9
5	3	15	5½	25	9½
6	3	16	6	26	9¾
7	3¼	17	6½	27	10
8	3¼	18	6½	28	10½
9	3¾	19	7	29	10½
10	3¾	20	7½	30	11
11	4¼	21	7½		
12	4½	22	8		

Multiply figure in inches by 2·54 to obtain metric equivalent in millimetres.

METHOD FOR FINDING THE WEIGHT OF MOLTEN METAL IN POURING LADLES

The diameter of crane ladles is usually greater at the top than the bottom. Thus the sides slope inwards with a uniform taper. As the angle of slope is not very great, the effective diameter may be taken as the average of the top and bottom diameters.

Thus the diagram may be taken to represent the internal dimensions of a ladle 300 mm diameter at the top and 200 mm diameter at the base. Therefore, the average or effective diameter is 250 mm, which is the diameter at half depth.

300 mm dia.

250 mm dia. at half depth

350 mm deep

200 mm dia.

To obtain the volume of the ladle, the average diameter is squared and multiplied by 0·7854, which gives the average cross-sectional area. This, in turn, is multiplied by the depth so as to get the volume. Keeping to the example illustrated·

250 x 250 x 0·7854 = 49,087 mm^2, say 490 cm^2
490 x 35 (ht in cm) = 17,180 cm^3 approximately

Different metals and alloys have different densities, the more commonly used being as follows·

Weight (g) per cm^3 of some metals and alloys (approximate)

Cast Iron	7·0	Aluminium	2·7
Copper	8·9	Steel	7·9
Lead	11·4	Brass and bronze	9·0
Magnesium	1·74		

Then if it is desired to know how much molten cast iron the ladle can hold, multiply its volume in cubic centimetres by the weight of one cubic centimetre of cast iron, thus:—

17,180 x 7·0 = 120,260 g

which is 120·26 kilograms.

If we take the average cross-sectional area of the ladle and a layer 1 cm deep, we get 490 cm^3, which is equivalent to a weight of 490 x 7·0 = 3·43 kg of cast iron. Therefore, we can say that the ladle holds approximately 3·43 kg of cast iron for each cm of depth.

It is possible, from this, to gauge the weight of any quantity of molten cast iron, in the ladle, simply by measuring its depth with an iron rod bent at right angles. For example, if the depth of liquid iron was found to be 13 cm, the weight of metal in the ladle would be 3·43 x 13 = 44·59 kg.

The same procedure is used for any other metal or alloy, but being careful to take its correct weight in grams per cubic centimetre as shown above.

GATING TERMS AND METHODS

Parting Gates

1 2 3 4	A

1 Sprue cup
2 Sprue
3 Skim bob
4 Gate (in-gate)

1 2 3 4 5 6	F

1 Pouring basin
2 Sprue
3 Strainer gate (top)
4 Strainer gate core
5 Strainer gate (bottom)
6 Gate

1 2 3 4 5	B

1 Sprue cup
2 Sprue
3 Gate
4 Relief sprue
5 Choke (in-gate)

1 2 3 4 5	G

1 Pouring basin
2 Splash core
3 Pouring box
4 Sprue
5 Gate

1 2 6 3 4 5	C

1 Sprue cup
2 Sprue
3 Gate
4 Choke
5 Gate
6 Skim bob (cope)
7 Skim bob (drag)

Bottom gate

1 2 3	H

1 Sprue cup
2 Sprue
3 Core gate

1 2 3 4 5	D

1 Pouring basin
2 Dam
3 Sprue
4 Gate
5 Choke

Branch gate

	I

1 Sprue
2 Relief sprue
3 Runner
4 Finger gates
5 Flat gate

1 2 3 4	E

1 Sprue cup
2 Sprue
3 Shrink bob
4 Gate

Pencil gate J

1 Pouring basin
2 Pencil gate

Wedge gate N

1 Wedge gate
2 Choke

Horn gate K

1 Horn sprue
2 Choke

Ring gate O

1 Sprue cup
2 Sprue
3 Ring gate

Top gate L

1 Pouring cup
2 Cup strainer core
3 Sprue

Whirl gate P

1 Sprue cup
2 Sprue
3 Runner
4 Pool
5 Relief sprue
6 Gate
7 Riser

Finger gate M

1 Sprue
2 Finger gate

The accompanying illustrations show the nomenclature recommended by a sub-committee of the American Foundryman's Society (AFS) for the various types of gating arrangements. It is recognised that very many local terms, colloquialisms, and variants are used, but the names given are largely self-explanatory and should be identifiable everywhere.

80

HOW TO MAKE
FEEDEX EXOTHERMIC SLEEVES

MIXING

FEEDEX exothermic material is mixed with water, either by hand, in a sand mill, or in a mixer until the consistency of a mixed core sand is obtained. Note that the green bond is somewhat lower than that of an average oil-sand mixture.

Water addition must be strictly controlled and varies with each type of FEEDEX as shown in the table below.

Type of FEEDEX	Optimum % Water	Pints Water per 25 lb.	Litres Water per kg.	Milling Time secs.	Optimum Drying Temperature
3	4	⅘	0·04	60-180	180-200°C. (350-390°F.)
4	5	1	0·05	60-180	180-200°C. (350-390°F.)
50	10	2	0·10	60-290	180-350°C. (350-650°F.)
93	10	2	0·10	180	180-200°C. (350-390°F.)

Note: Once water has been added FEEDEX should be moulded and dried within two or three hours.

Feeding sleeve

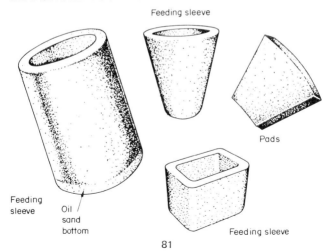

Pads

Feeding sleeve

Oil sand bottom

Feeding sleeve

MOULDING

The mixed FEEDEX is moulded into sleeves or required shapes as shown on page 81. A layer of core sand 13 mm thick is incorporated at the end of the sleeves which will be in contact with the casting, otherwise the sleeve may "burn on" to the casting.

VENTING

Venting is necessary to exhaust the gases of combustion and allow free access of air to complete the exothermic reaction. Sleeves should be vented to within 10 mm of the bottom and it is also important to vent the backing sand. This is clearly illustrated in the drawings shown on this page.

DRYING

Drying temperature of FEEDEX exothermic materials 3 and 4 should not exceed 200°C (390°F) otherwise there is a risk of premature ignition. Drying time depends largely on the local conditions, a fair guide is to dry for one hour per 25 mm wall thickness. FEEDEX 50 may be dried at between 180°C (360°F) and 350°C (660°F). Due to the increased water content of this material the drying time must be increased by 50%.

RAMMING FEEDEX SHAPES
IN SAND MOULDS

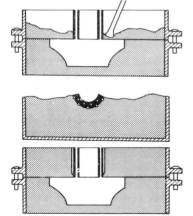

Ramming sand in cope in which the FEEDEX sleeve is positioned.

Rammed mould showing FEEDEX sleeve in position and vent holes both in sleeve and surrounding sand.

PRACTICAL GUIDE TO THE DETERMINATION
OF FEEDER SLEEVE DIMENSIONS

The following is a very general guide and deals only with open top feeders. See also "The application of Mouldable Exothermic Materials to Steel Castings" in Section VI.

The same calculations can and should be employed for determining feeder sleeve dimensions when using the alternative fully insulating (KALMIN) or self-heating refractory sleeves (KALMINEX). These newer materials are more effective for some applications than a fully exothermic sleeve and are far more resilient and able to withstand rough use. Because of their extremely high short- and long-term insulation value loss of heat from the open top of a riser assumes increased importance and all sleeved heads should be covered with a suitable hot topping or anti-piping compound immediately on completion of pouring.

STAGE I

To determine the sleeve dimensions required for a particular section of casting which is to be fed, the type of alloy and ruling section of the casting are the major considerations.

D = ruling section representing part of casting to be fed. This dimension determines FEEDEX sleeve diameter. (See table below.)

Alloy	Internal Sleeve Diameter
Steel	D or greater than D★
High Duty Irons	$\frac{2}{3}$D to D★
Malleable Iron	$\frac{2}{3}$D to D★
Ductile Iron	$\frac{2}{3}$D to D★
Grey Iron	$\frac{1}{2}$D to $\frac{2}{3}$D★
Copper Base Alloys	$\frac{2}{3}$D to D★
Light Alloys	$\frac{1}{2}$D to D★

★When the section to be fed is deep or has sections attached to it, the internal sleeve diameter should be greater than D or approaching the upper limit of the recommendations given above.

Note on Sleeves for Grey Iron

FEEDEX sleeves with an integral breaker core are preferred for grey iron castings. This minimises the formation of a coarse structure beneath the riser—a disadvantage which may result from the use of straightforward cylindrical sleeves. Separate oil sand breaker cores, used with a cylindrical sleeve can be used if desired. To obtain correct breaker core dimensions or given size of sleeve calculate as follows:—

Breaker core orifice diameter — divide internal sleeve diameter by 2·5.

Breaker core thickness — divide internal sleeve diameter by 10.

STAGE 2

Next determine sleeve height. This should ideally be from 1—1·5 times the internal sleeve diameter. FEEDEX sleeve height is very often influenced by depth of moulding boxes, heights of runner bushes, etc. In practice, therefore, FEEDEX sleeve heights are often from 0·8—2·0 times their diameter.

STAGE 3

From the formula 3·14 x r^2 x h, where r is sleeve radius and h sleeve height, calculate the internal volume of the sleeve. The resulting volume should approximate to the percentage quantity of liquid metal required to adequately feed that cast section to which the sleeve is attached.

Volume of feed metal required for various alloys

Steel	15% to 25%
High Duty Irons ...	12% to 17%
Malleable Iron ...	12% to 17%
Ductile Iron	12% to 17%
Grey Iron	Approx. 6%
Copper Base Alloys ...	12% to 17%
Light Alloys	10% to 15%

STAGE 4

Selecting the right Type of Riser Sleeve Material
    FEEDEX exothermic materials for sleeve making are available in several grades so that ignition sensitivity and burning rate are suited to all types of alloy.

    The following table gives principal recommendations taking into consideration sleeve diameters. Other types of FEEDEX material are available for special applications.

Table 3

Alloy	FEEDEX 3 (mm)	FEEDEX 4 (mm)	FEEDEX 50 (mm)	FEEDEX 93 (mm)
Steel	50 to 150	Up to 50	Over 150	50 to 150
Iron (all types)	50 to 200	,, ,, 50	,, 200	50 to 200
Copper base alloys	50 to 200	,, ,, 50	,, 200	50 to 200
Light alloys	Over 100	,, ,,100	—	—

FEEDEX 3 and 93 are frequently interchangeable. Steel foundries generally require a clean strip on heads so FEEDEX 3 is preferred.

Determining Sleeve Thickness
    Refer to the following table for determining sleeve thickness requirements.

English		Metric	
Internal Diameter (in.)	Wall Thickness (in.)	Internal Diameter (mm)	Wall Thickness (mm)
Below 3	$\frac{1}{2}$	75 or less	12
4–6	$\frac{3}{4}$	100–150	18
7–9	1	175–225	25
10–12	$1\frac{1}{4}$	225–305	31
13–15	$1\frac{1}{2}$	330–380	37
16–18	$1\frac{3}{4}$	405–455	43
19–21	2	480–535	50
22–24	$2\frac{1}{4}$	560–610	56
25–27	$2\frac{1}{2}$	635–685	62
28–30	$2\frac{3}{4}$	710–760	68

Use the table below to determine the type of insulating material required for a non-exothermic application.

Alloy	Riser Diameter Up to 125 mm	Riser Diameter Above 125 mm
Aluminium	KALMIN 33	KALMIN 33
Copper base	KALMIN 44	KALMIN 44
Cast iron	KALMIN 4170	KALMINEX 30
Steel	KALMIN 4170	KALMINEX 30

SAND TEST DATA WHICH HAVE PROVED SATISFACTORY FOR VARIOUS CASTINGS IN DIFFERENT CASTING ALLOYS

The test results recorded were obtained using the American Foundrymen's Association apparatus and are after H. Dietert.

Casting Alloy	Moisture %	Permeability	Green Compression lbs. per sq. in.	Clay Substance %	Fineness No. (see page 97)
Aluminium Castings	6·5 to 8·5	7 to 13	6·5 to 7·5	12 to 18	225 to 160
Brass and Bronze Castings ...	6·0 to 8·0	13 to 20	7·0 to 8·0	12 to 14	150 to 140
Copper-Nickel Castings ...	6·0 to 7·5	37 to 50	6·5 to 8·0	12 to 14	130 to 120
Grey Iron, light stove plate castings	6·5 to 8·5	10 to 15	6·0 to 7·5	10 to 12	200 to 180
Grey Iron, squeeze moulds ...	6·0 to 7·5	18 to 25	6·2 to 7·5	12 to 14	120 to 87
Grey Iron, medium floor sand	5·5 to 7·0	40 to 60	7·5 to 8·0	11 to 14	86 to 70
Grey Iron, medium, synthetic sand	4·0 to 6·0	50 to 80	7·5 to 8·5	4 to 10	75 to 55
Grey Iron, heavy, green or dry sand	4·0 to 6·5	80 to 120	5·0 to 7·5	8 to 13	61 to 50
Malleable Iron Castings, light	6·0 to 8·0	20 to 30	6·5 to 7·5	8 to 13	120 to 92
Malleable Iron Castings, heavy	5·5 to 7·5	40 to 60	6·5 to 7·5	8 to 13	85 to 70
Steel, light green sand castings	2·0 to 4·0	125 to 200	6·5 to 7·5	4 to 10	56 to 45
Steel, heavy green sand ...	2·0 to 4·0	130 to 300	6·5 to 7·5	4 to 10	62 to 38
Steel Castings, dry sand ...	4·0 to 6·0	100 to 200	6·5 to 7·5	6 to 12	60 to 45

GREEN SAND FACING MIXTURES FOR IRON CASTINGS USING NATURALLY BONDED SANDS

Casting Section	New Sand	Floor Sand	Coal Dust
up to 6 mm	6 parts	20 parts	1½ parts
6—13 mm	8 ,,	20 ,,	1¾ ,,
13—18 mm	10 ,,	20 ,,	2½ ,,
18—40 mm	15 ,,	20 ,,	3 ,,
above 40 mm	20 ,,	20 ,,	4 ,,

The above mixtures need to be well mixed or lightly ground in an edge runner type sand mill and used to cover the pattern. In each case the mould can be finished off by dusting with the TERRA-POWDER dressing which can be sleeked if required.

"New Sand" in the mixtures refers to any of the well known naturally bonded sands. In each case the mixture can be applied equally well to non-ferrous casting by omitting the coal dust and replacing it by TERRADUST dressing.

In the case of moulds for non-ferrous work the correct type of dusting dressing is TERRACOTE or TERRADUST.

All facing sand should be used as dry as possible consistent with good working. An average moisture content is about 6%.

SYNTHETIC SAND MIXTURES FOR IRON CASTINGS

Green Sand Facing (proportions by weight).

  60% Old silica sand.
  34% to 37% New silica sand.
  1% to 4% Bentonite (dependent on green strength required).
  2% Coal dust.

Core Sand for Average Castings (proportions by weight).

  93% Silica sand.
  5% Core oil.
  2% Bentonite.

Core Sand for Heavy Section Castings (proportions by weight).

  83% Silica sand.
  9·5% Silica flour.
  5·0% Core oil.
  2·5% Bentonite.

SAND ADDITIVES FOR IRON CASTINGS

Traditionally most iron castings have been made in natural, semi-synthetic or synthetic green sand to which a very variable addition of coal dust (2-10%) has been made. The inclusion of coal dust in the mix ensures good strip of the sand from the casting and a better surface finish. Opinions differ as to how this is accomplished, but some features are probably important. Combustion of some of the coal dust as the metal enters may ensure oxygen-free reducing conditions in the mould cavity, thus reducing oxide formation and aiding apparent fluidity. Evolution of gas from the heated coal dust can also be said to produce a gas cushioning effect, thus giving finish and polish to the casting. Strong heating of coal dust also ensures the formation and deposition of a type of carbon layer, known as lustrous carbon, which is not wetted by metal and so produces improved surface finish in this way. Certainly the inclusion of certain types of volatile material in the sand can ensure castings with better strip and surface finish and can also materially help to eliminate sand expansion defects such as rat-tails, buckles, scabs, etc. Unfortunately, coal dust is extremely dirty to work with, and mill and shake-out areas can be very dusty indeed· the mixed sand is also dirty to handle.

Many if not all of these objections can be removed by the alternative use of specially blended additives such as BENTOKOL. Where the clay/volatile ratio in the sand is suitable, one or another of the BENTOKOL versions can be substituted for the coal dust. A complete substitution can often be made, but balancing additions of supplementary clay or volatile can be introduced if necessary. The use of BENTOKOL often means that a "one-shot" sand addition is sufficient and always improves working conditions. Fume after casting is reduced, and the whole foundry is cleaner and pleasanter while casting finish is maintained or often enhanced.

BENTOKOL is basically a blend of natural clays to which have been added balanced essential volatiles to produce a light coloured, virtually dust-free product. In addition, a number of other, powerful, ingredients are present to provide special properties having a far-reaching effect on the chemistry of the whole clay/volatile system. The result is a composite green sand conditioner which provides both bond and volatile content for iron foundry moulding sands, often eliminating the need for other additives.

EXAMPLE OF ROUTINE SAND CONTROL TESTS

Sample of Typical Mansfield Moulding Sand (unmilled)

Moisture	5·0 to 7·0%
Permeability (A.F.S.)	25·0 to 30·0%
Green Compression Strength (A.F.S.) ...	4·5 to 6·0 lbs. per square inch
Clay substance	10·0 to 12·0%

Sieve Grading — British Standard Sieves (B.S.S. 410-31).

Sieve No.				Per Cent retained
8	...	...	...	0·04
10	...	...	...	0·03
16	...	...	...	0·10
22	...	...	...	0·22
30	...	...	...	0·55
44	...	...	...	2·23
60	...	...	...	2·22
100	...	...	...	10·02
150	...	...	...	34·70
200	...	...	...	28·35
300	...	...	...	7·31
Pan	...	...	...	4·05
Clay Substance		...		10·18

Interpretation of Sand Test Results

Although no hard and fast rules can be laid down with respect to sand properties and grain scatter, there are general guiding principles.

Excessive moisture should be avoided, as this will tend to lower the permeability and increase the possibility of blown castings. Plasticity and deformation of a sand also increase with the moisture content. Low permeability and compression strength encourage the entrapment of gas and the washing away of sand by the flow of the molten metal.

The free venting properties or the permeability of a sand will depend largely on the distribution of the sand grains on different sieves. The less scattered the grains, the higher the permeability. Therefore, maximum permeability would be obtained in sand having all the grains of the same size, that is of a single sieve grading.

In practice, it is possible to maintain a distribution where 60% to 80% of the sand grains are retained on three adjacent sieves. If the sand were composed entirely of rounded grains of uniform grading it would most likely give trouble due to scabbing and if the grains were fairly coarse, metal penetration could be expected.

Fines or silt in a moulding or core sand may be regarded as all material passing the 200 mesh sieve. Burnt out clay, coal dust and ruptured quartz grains help to form silt. Too much silt seriously lowers the permeability. On the other hand, some silt is of advantage in strengthening the bond and preventing metal penetration and scab formation. Apart from the clay substance, which has a grading equal to silt, about 5% of minus 200 mesh grade is about right.

The test values given for the specimen sand, although they show some range of properties, should be taken as a general example only. Each type of casting requires a sand with properties which vary according to the metal or alloy concerned and to the points most desired in the cast components. For example, decorative castings need fine detail and good as-cast appearance. Fine grained sands must therefore be employed, and green compression strength is obtained at the expense of permeability.

Alternatively, heavy sectioned castings call for high permeability sands. Values between 40 and 60 or even higher may be specified. Such "open" sands tend to possess a low green compression strength which may have to be made up by using additions such as bentonite.

Sands with a high silt fraction are often not workable with moisture contents of less than 7%. Coarse, open sands with grains mainly contained on the larger mesh sieves will bond with less moisture. Incidentally, they also tend to dry out more rapidly in use, and are not so easily patched, in the case of hand moulding.

ART CASTINGS

For art castings produced by sand moulding methods, a very fine grade of sand is desirable, especially in the case of small work and where detail is very intricate, a natural clay bonded sand such as Erith is suitable, but a synthetic mixture based on Ryarsh silica sand is appended :—

150 kg of Ryarsh silica sand.
2.75 kg of bentonite.
2.75 kg of Terradust No. 2.
0.9 kg of coregum powder.
4 litres of water.

First, mix all the ingredients together dry, and then add the water gradually and finish milling. The milled sand is put through an aerator.

COMPARISON OF STANDARD SIEVES

British Standard Sieve No.	American Society for Testing Materials Sieve No.	Tyler Sieve No.	German Din 1171 Sieve No.	B.S.S. Aperture Diameter	A.S.T.M. Aperture Diameter	Tyler Aperture Diameter	Din 1171 Aperture Diameter
8	—	9	—	2·06 mm.	—	2·00 mm.	—
10	12	10	4	1·68 mm.	1·68 mm.	1·68 mm.	1·5 mm.
16	18	16	6	1·00 mm.	1·00 mm.	1·00 mm.	1·0 mm.
22	25	24	8	0·70 mm.	0·71 mm.	0·71 mm.	0·75 mm.
30	35	32	12	0·50 mm.	0·50 mm.	0·50 mm.	0·5 mm.
44	45	42	16	0·35 mm.	0·35 mm.	0·35 mm.	0·4 mm.
60	60	60	24	0·25 mm.	0·25 mm.	0·25 mm.	0·25 mm.
100	100	100	40	0·15 mm.	0·15 mm.	0·15 mm.	0·15 mm.
150	140	170	60	0·10 mm.	0·10 mm.	0·09 mm.	0·10 mm.
200	170	200	80	0·08 mm.	0·09 mm.	0·08 mm.	0·08 mm.
300	270	270	E 110	0·05 mm.	0·05 mm.	0·05 mm.	0·05 mm.
350	400	—	E 185	0·03 mm.	0·04 mm.	—	0·03 mm.

CORE SAND MIXTURES

The products described in this section are marketed by the FOSECO organisation throughout the world, excluding the United Kingdom and France.

Foseco Fordath Products

There are three main groups of core binders in general use, namely: —

Group 1. Oil and cereal.
Group 2. Oil, dextrin and water emulsions and mixtures.
Group 3. Synthetic resins.

Both the oil and cereal (added separately plus water) and the oil, dextrin, and water emulsions and mixtures, are used throughout the industry, the oil imparting the dry strength and the cereal or dextrin giving the green strength to the core sand mix.

The synthetic resins are now widely established in the industry.

Groups 1 and 2

The following table illustrates the formulation of various types of core sand mixes from the above groups.

Mixture No.	Alloy	General Application and Description of Core Sand Properties	Dry Silica Sand	Moist Naturally Clay bonded sand	Glyso Semi-Solid	Glyso Cream	Permol	X.L. Cereal	Water
			kg	kg	kg	kg	kg	kg	ml
1	Grey Iron	General Castings. Strong green bond. Medium dry strength	40	10	1.5				
2	Grey Iron	General purpose blowing or shooting core sand mixture. Low green strength	50			1.25			
3	Grey Iron	Good quality blowing or shooting core sand mixture. Medium green and high dry strength.	50				0.5	0.5	200–250
4	Steel (light section)	Suitable for blowing or shooting. Medium green and high dry strength	50			1.25–1.5			
5	Steel (medium section)	A good general purpose mix. Higher green, dry, and hot strengths than above	38.75	11.25		1.6			
6	Light Alloy	For aluminium and magnesium castings, etc. Good green strength coupled with sufficient dry strength	50		1.25				35–70
7	Copper base alloys	A first class core sand mix offering a high green bond and hot strength	45	5	1.25		0.5		

Group 3

The main types being used are·
    (1) Urea formaldehyde/furfuryl alcohol.
    (2) Phenol formaldehyde/furfuryl alcohol.
    (3) Phenol urea formaldehyde/furfuryl alcohol.
    (4) Phenol formaldehyde Novolak resin (shell moulding).

COLD SET   For Production of Medium and Large Cores

(1) UF/FA. Fursatils CS 150, 120, 90. Excellent high quality general purpose resins suitable for the production of cores and moulds in both ferrous and non-ferrous foundries.

(2) PF/UF/FA. Fursatil CSFM 22. A low nitrogen level (2.5%) binder developed as a unit sand resin for moulding and core making applications. Considerable protection against nitrogen pin-holding is afforded without the problems of brittle bond sometimes experienced with low nitrogen and nitrogen free resins.

(3) PF/FA. Fursatil CS 300, CS 1003. Completely nitrogen free resins developed specifically for use in steel and high duty iron. foundries and for sand reclamation where maximum protection against pin-hole porosity is required.

HOT BOX for High-Speed Production of Repetition Cores.

(1) UF/FA. Fursatil HB 21 and HB 27. High-quality furane hot box resins suitable for a wide range of core making applications in both ferrous and non ferrous foundries. They give fast through-cure and high core strengths. The mixed sand has very good flowability allowing complex cores to be blown without difficulty.

(2) PF/FA. Fursatil. HB 31. Specially developed for high duty iron, SG iron and steel to give maximum protection against pin-hole porosity. HB 31 has excellent core-making characteristics. It uses a two-catalyst system which ensures low fume evolution, excellent core finish, and surface hardness.

SHELL MOULDING

Phenol formaldehyde condensates (Novolaks) formulated for use in the manufacture of resin-coated sand for the shell process (Croning Process).

Resins for Warm Coating:— Grades 777, 728, 728/S. Liquid resins giving optimum all-round strength and cure characteristics.

Resins for Hot Coating:— FPN 96, FPN 100. Solid resins in flake form giving high strengths and investment rates.

Grades 109A, 109B. Liquid resins for use where equipment requirements make it preferable to dispense a liquid chemically equivalent to the solid resins.

Note: Product data sheets giving full technical and mixing instructions are available on all the above products.

PLASTSAND

A range of pre-coated sands suitable for all types of ferrous, non-ferrous and steel shell core and mould processes. Resin contents vary from 0.8-6.5% dependent on application.

Notes on the Application of Fursatil

(a) Hot Box Resins

(i) *Amount to use* varies between 1·5-2·5%, depending on core strength required, cleanliness, and fineness of sand.

(ii) *Mixing procedure* (1). Clean mixer. (2). Charge cold dry sand into the mixer.
(3). Add catalyst and mix for 1 minute.
(4). Add resin and mix for 2-3 minutes.

(iii) *Bench life of sand*. Six hours under normal production conditions.

(iv) *Core production*. Blowers or shooters and metal core boxes heated between 450-500°F (230-260°C) are normally used. Cores are cured ready for ejection usually between 8-25 seconds.

(b) Cold Set Resins

(i) *Amount to use*. This varies between 0·8-2·5% depending on the core strength required, cleanliness, and fineness of sand.

(ii) *Mixing Procedure*. (1). Clean mixer
(2) Charge cold dry sand into the mixer.
(3) Add catalyst (30-50% of the weight of resin) and mix for 1-2 minutes.
(4) Add resin and mix for 2-3 minutes.

(iii) *Core making*. The core sand mix has extremely low green strength and good flowability, consequently requiring hardly any ramming. Stripping can be carried out after the core has hardened. This is usually between 30-60 minutes, depending on the character of the sand mix and the size of the core.

Note: Fully automatic mixers are now widely used for the production of synthetic resin cold set and hot box core sand mixes.

THE CARBON DIOXIDE PROCESS

For Hardening Sand Moulds

The CO_2 process, as it is generally known, consists of mixing a clean dry silica sand with a silicate binder, compacting the mixture to shape and hardening it by passing carbon dioxide gas. Moulds and cores prepared by this process are ready for immediate use.

Choice of Sand

Silica sands as free from clay as possible must be used and the sand must be cold and dry. A maximum moisture content of 0·5% is permissible.

CO_2 Binder Content

The quantity of silicate binder required depends upon the average grain size or fineness number of the sand. The graph below shows the binder requirements for a wide range of sand fineness values, using CARSIL. Note that as the fineness number increases a greater percentage of binder addition is required. Refer to following page for method of calculating the fineness number of a sand.

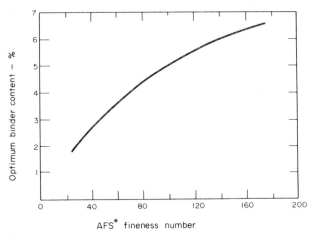

*American Foundrymen's Society.

HOW TO
CALCULATE SAND FINENESS NUMBER

The equipment required is a nest of standard laboratory sieves, a sieve vibrating mechanism and a sensitive balance accurate to plus or minus 0·5 g.

Test Procedure

1. Weigh a 100 g. sample of the sand when it is perfectly dry.
2. Place the entire sample into the top (coarsest) sieve, place nest of sieves on vibrator, vibrating for 15 minutes.
3. Remove sieves, and, beginning with the top sieve, weigh the quantity of the sample remaining on each sieve.
4. Arrange the respective weights in a column as shown below, adjacent to their sieve mesh number (either the British standard sieve numbers or the A.F.S. American standard sieves may be used).
5. Multiply each separate sieve weight by the **preceding** sieve mesh number.
6. Divide the total product by the total sample weight and this produces the Fineness Number, which is the A.F.S. Fineness Number when calculating by this method.

Sieve No. British Standard	Sand Retained on Sieve (g.)	Multiplied by Previous Sieve No.	Product
+10	Nil.	—	—
+16	Nil.	—	—
+22	0·2	0·2 × 16	3·2
+30	0·8	0·8 × 22	17·6
+44	6·7	6·7 × 30	201·0
+60	22·6	22·6 × 44	1104·4
+100	48·3	48·3 × 60	2898·0
+150	15·6	15·6 × 100	1560·0
+200	1·8	1·8 × 150	270·0
Through 200	4·0	4·0 × 200	800·0
		Product =	6854·2

$$\text{A.F.S. Fineness No.} = \frac{\text{Product}}{\text{Weight of Sample,}} = \frac{6854 \cdot 2}{100} \text{ or,}$$

Which is 68 A.F.S.

GENERAL AND PRACTICAL DATA ON THE CARBON DIOXIDE PROCESS

Collapsibility of Moulds and Cores

Breakdown characteristics after casting can show considerable variation depending on the temperature to which the sand has been heated. In most cases and particularly where the ratio of sand to casting is high, some form of additive is necessary to assist collapsibility at knock-out.

In general, breakdown additives act as binder extenders and also provide a controlled source of volatile materal preventing fusion of sand grains at high temperatures. Several types are available depending on the particular properties required. The group name is DEXIL (followed by an identifying number) and ½-1% by weight would be a typical addition rate.

Effect of Clay

As little as 0·5% clay in the sand is capable of reducing the hardened strength of CO_2 bonded cores, even after a short standing period. However, if it is essential to have high green strength then up to 1% of Ball Clay may be added. Clay of the bentonite type should not be used.

Surface Finish

Ferrous castings, and heavy castings of any alloy produced in silicate bonded moulds may show a rough surface due to metal penetration. The application of mould or core dressings is therefore essential and spirit based coatings, which dry on igniting, are most suitable.

Recommended Coatings

For light alloys	MOLDCOTE 9 or MOLDCOTE 3 (self-drying).
Copper base alloys:	MOLDCOTE 6 or 10.
Cast iron:	MOLDCOTE 6 or 11.
Steel:	MOLDCOTE 33 or CERAMOL 55.
Manganese steel:	MOLDCOTE 31.

In addition to the above, traditional type dressings others are available in a ready-to-use form.

MOLCO 277, equivalent to MOLDCOTE 10 or 11
MOLCO 268, equivalent to MOLDCOTE 6.
CERAMOL 256, equivalent to CERAMOL 55.

GASSING TECHNIQUES FOR THE
CARBON DIOXIDE PROCESS

Gassing Time

Practical experience shows that it is better to under-gas a mould or core rather than to over-gas. Over-gassing is manifest by the appearance of white sodium bicarbonate crystals on the mould or core surface. It also signifies that the surface will be friable and erode readily by the action of molten metal during pouring.

Applying CO_2 Gas

Methods of introducing the gas vary from a single tube or manifold, to covers or hollow patterns and pattern plates. Gassing from underneath a mould appears to give the most satisfactory results and should be used wherever possible. Some methods of application are shown on the following page.

Where the CO_2 gas is introduced by way of tubes attached to a hood or cover board, these covers normally consist of sheets of plywood or hardboard edged with a wooden strip and faced with soft rubber. The rubber seal around the moulding box ensures rapid gassing times and the minimum wastage of CO_2 gas.

Core Boxes

These should be designed so that air contained within the sand mass can be readily displaced when introducing CO_2 gas. Prior to gassing, the core boxes should be sharply rapped to allow easy stripping after hardening. Some typical methods of core production by the CO_2 process are shown on the following page.

Gas Consumption and Pressure

Gas consumption will vary from 1% to 2% CO_2 of the weight of moulds or cores to be gassed. Gas pressures are generally low, within the range of 15 to 20 lb per square inch (1 to 1·4 kg/cm^2). At these pressures, gassing times of 15-30 seconds are common.

METHODS OF GAS APPLICATION

TYPICAL METHODS OF CORE PRODUCTION BY CO₂ PROCESS

Tube and hood methods

Gassing hood method

Gassing chamber

Hollow core method

ALTERNATIVE METHODS OF HARDENING SILICATE BONDED SAND

In addition to carbon dioxide gas, other materials have the property of inducing a set in the types of sodium silicate normally used for foundry purposes. These can include, for example, anhydrite, some forms of cement, dicalcium silicate, finely ground ferro-silicon (the Nishiyama Process), ground blast furnace slags, etc. Most of these materials have limitations, through-curing is often difficult to obtain as is close control over the setting time, while some (the N Process) evolve hydrogen during the reaction and free venting must be available to prevent the possibility of an explosive gas build-up. Heat is frequently evolved also and care must be exercised in the selection of pattern materials and paints. All these additives are finely ground powders normally added to the mill before the silicate. They do therefore tend to dust unpleasantly.

Practically all these objections can be overcome by using the liquid ester type of hardening agent. Being liquids these are easily handled, particularly in conjunction with screw type continuous mixers, and a wide range of agents is available so that almost any desired setting time can be obtained. The setting process is temperature dependent, but once the sand temperature at discharge is known and the setting time required has been decided, the appropriate product from the range available can be selected. Results are then consistent and reproducible.

The CARSET range of setting agents includes materials giving setting times of between 10-120 minutes, the exact time for any particular agent being, as already mentioned, dependent on the mix temperature. The range is colour coded as well as being numerically identified, so that the particular setting agent in use can be checked at a glance. More detailed using instructions including operating graphs can be obtained.

While the CARSET range can be used in conjunction with straight foundry grade silicates of the 112 type, the best results are obtained with special high ratio types such as 100. Even better results, especially where good breakdown is necessary, are obtained with a proprietary silicate such as CARSIL 100 which contains special additives. Normal breakdown additives such as those added as powders or liquids to a conventional CO_2 process mix cannot be used with the self-setting process.

Advantages associated with a self-setting silicate process may be summarised as follows:—

1. Total binder and catalyst costs are significantly lower than those of conventional air and cold setting resins and oils.

2. Mixing may be performed satisfactorily in either batch or continuous mixers.
3. Mixer and equipment clean-down is facilitated by the water solubility of the uncured system.
4. The mixed sand is free from objectionable odour. Dermatitic and toxic hazards to operators are negligible. No hydrogen evolved during setting.
5. The desired bench life and stripping time can be accurately achieved and controlled.
6. Complete through-cure is obtained.
7. No heat is generated during curing and thus there is no expansion to cause problems of dimensional accuracy after pattern removal.
8. Gas evolution during casting is small and no unpleasant fumes are evolved.
9. A nitrogen-free system eliminates trouble from subsurface pinholes in iron castings.
10. The incidence of finning and metal penetration defects is minimised.
11. The storage life of moulds and cores is excellent.
12. Moulds and cores are dimensionally stable during casting; therefore, minimal dilation and more accurate reproduction of pattern dimensions in the castings is possible.
13. Casting strip is excellent, easy shake-out after casting is obtainable and the used sand may be easily re-claimed.
14. Since the system is carbonate-free, contamination of green sand is not detrimental.
15. It is possible to incorporate low cost, shell-containing sea sand with this process if other considerations permit.

LINING MIXTURES FOR POURING LADLES AND HAND SHANKS

Crane Ladle Lining (proportions by weight).

70% Amorphous crushed silica (ganister rock).
30% Fireclay.

Add water to give required working consistency when milled. Additional water is added if the mixture is to be used as a brick jointing or facing.

Pouring Ladles and Hand Shanks.

Synthetic Mix.

43% by weight new silica sand.
43% by weight old silica sand (e.g., old core sand).
14% by weight bentonite.

Water to give consistency of wet moulding sand after milling.

Natural Mix.

50% by weight of new red sand (Worksop, Mansfield, etc.).
50% by weight of ganister mixture as used for cupola patching.

Both the above mixtures should be thoroughly milled in an edge runner type mill.

The steel ladle shells to be lined are first given a coating of wash made by diluting the lining mixture so that it can be painted on. Then straw rope is placed across the base of the ladle shell and up the sides to act as a vent for the escape of steam when drying the lining. For hand shanks, two lengths of straw rope run across the bottom and up the sides at right angles to each other, will be sufficient. Large ladles may need rather more straw rope, but in any case it is usual to drill a number of 5 mm diameter holes in the base of large ladles to act as vents for the escape of steam. After lining, the ladles should be given a thorough drying, finally at a red heat.

The lining mixture, after milling with water to develop the requisite degree of plasticity is placed in position in the bottom of the ladle by hand and then extended up the side walls. It is smoothed off with a water brush prior to drying as described. Thorough drying of the lining is most important. Sometimes the lined ladles are given a coating of plumbago or similar black wash when they are appreciably warm. This helps in keeping them clean and free from adhering dross and "build up."

BOTTING PRACTICE
FOR CUPOLA OPERATORS

The British Cast Iron Research Association, in their publication, Broadsheet No. 3, state that a refractory mixture, based on fireclay is the most satisfactory botting material.

A botting mixture should combine the qualities of high fusion temperature, medium dry strength and low drying contraction. The B.C.I.R.A. report that the following mixtures give satisfaction in practice.

TYPE A
Suitable for small botts and frequent tappings.

Fireclay (30–50% on clay grade)—70–88%.
Coat Dust (11% ash max., superfine grade)—10–20%.
Note : 2–10% medium fine sawdust can be used instead of coal dust if required.

TYPE B
Suitable where long gathering periods and large tap hole is employed.

Fireclay (as above)—50%.
Coal dust or sawdust (as above)—10–20%.
Black sand—20–38%.

Design of Tap Hole

Of the various forms of tap hole used experience shows that the design shown below gives the most consistently good results. With this design the dried plug can often be teased from the tap hole so that it is flushed away in one piece by the molten metal.

BOTTING PRACTICE

The mixtures are best mixed dry and water should be added gradually with thorough kneading. It is best to prepare the mixture the day before use, keeping it covered with a damp sack.

After making a conical shaped bott by hand it should be dipped in coal dust. This prevents "splutter" when the bott comes into contact with the metal stream.

The sketch below illustrates some useful features associated with good botting practice. Note the short bott stick which is easier to handle than a long one. The "splutter" shield as shown affixed to the bott stick is an added safeguard. Note also the bracket which holds spare bott sticks and the container of coal dust which is readily available. The bott stick should have a convex surface as shown. This is found to provide a more secure adhesion of the clay bott than a flat metal face.

SECTION II

LIGHT CASTING ALLOYS

ALUMINIUM CASTING ALLOYS — LM SERIES

Uses and General Remarks

LM0 Mainly used for sand castings for electrical, chemical, and food applications.

LM2 One of the two most widely used alloys for all types of die castings.

LM4 The most versatile of the alloys; has very good casting characteristics and is used for a very wide range of applications. Strength and hardness can be greatly increased by heat treatment.

LM5 Suitable for sand and chill castings requiring maximum resistance to corrosion, e.g. marine applications.

LM6 Suitable for large, intricate, and thin-walled castings in all types of moulds; also used where corrosion resistance or ductility is required.

LM9 Used for applications especially low-pressure die castings requiring the characteristics of LM6 but higher tensile properties following heat-treatment.

LM10 Mainly used for sand and chill castings requiring high strength and shock resistance. Requires special foundry technique; heat-treated.

LM12 Mainly used where a very good machined surface finish and hardness is required.

LM13 Mainly used for pistons.

LM16 Suitable where high mechanical properties are desired in fairly intricate sand or chill castings. Requires heat-treatment.

LM18 Combines good foundry characteristics with high resistance to corrosion.

LM20 Mainly used for die castings. Similar to LM6 but a little better machinability and hardness.

LM21 Generally similar to LM4-M in characteristics and applications but better machinability and higher proof strength.

LM22 Used for chill castings requiring good foundry characteristics with good ductility. Requires heat treatment.

LM24 One of the two most widely used alloys for all types of die castings.

LM25 Suitable where good resistance to corrosion combined with high strength is required.

LM26 Mainly used for pistons as alternative to LM13.

LM27 A versatile sand and chill casting alloy introduced as an alternative to LM4 and LM21.

LM28 Piston alloy with lower coefficient of expansion than LM13. Requires special foundry technique.
LM29 As LM28 but lower coefficient of expansion.
LM30 For unlined die-cast cylinder blocks with low expansion and excellent wear resistance.

The tables appearing on pages 110 to 120 inclusive are based on data given in The Properties and Characteristics of Aluminium Casting Alloys, published by the Association of Light Alloy Refiners and Smelters (Alar), London.

ALUMINIUM CASTING ALLOYS — LM SERIES

Chemical composition (%)[a]

BS 1490	Cu	Mg	Si	Fe	Mn	Ni	Zn	Pb	Sn	Ti	Others
LM0	0·03	0·03	0·30	0·40	0·03	0·03	0·07	0·03	0·03	–	Al 99·50 min
LM2	0·7-2·5	0·30	9·0-11·5	1·0	0·5	0·5	2·0	0·3	0·2	0·2	–
LM4	2·0-4·0	0·15	4·0-6·0	0·8	0·2-0·6	0·3	0·5	0·1	0·1	0·2	–
LM5	0·1	3·0-6·0	0·3	0·6	0·3-0·7	0·1	0·1	0·05	0·05	0·2	–
LM6	0·1	0·10	10·0-13·0	0·6	0·5	0·1	0·1	0·1	0·05	0·2	–
LM9	0·1	0·2-0·6	10·0-13·0	0·6	0·3-0·7	0·1	0·1	0·1	0·05	0·2	–
LM10	0·1	9·5-11·0	0·25	0·35	0·10	0·10	0·10	0·05	0·05	0·2[b]	–
LM12	9·0-11·0	0·2-0·4	2·5	1·0	0·6	0·5	0·8	0·1	0·1	0·2	–
LM13	0·7-1·5	0·8-1·5	10·0-12·0	1·0	0·5	1·5	0·5	0·1	0·1	0·2	–
LM16	1·0-1·5	0·4-0·6	4·5-5·5	0·6	0·5	0·25	0·1	0·1	0·05	0·2[b]	–
LM18	0·1	0·10	4·5-6·0	0·6	0·5	0·1	0·1	0·1	0·05	0·2	–
LM20	0·4	0·2	10·0-13·0	1·0	0·5	0·1	0·2	0·1	0·1	0·2	–
LM21	3·0-5·0	0·1-0·3	5·0-7·0	1·0	0·2-0·6	0·3	2·0	0·2	0·1	0·2	–
LM22	2·8-3·8	0·05	4·0-6·0	0·6	0·2-0·6	0·15	0·15	0·1	0·05	0·2	–
LM24	3·0-4·0	0·1	7·5-9·5	1·3	0·5	0·5	3·0	0·3	0·2	0·2	–
LM25	0·1	0·20-0·45	6·5-7·5	0·5	0·3	0·1	0·1	0·1	0·05	0·2[b]	–
LM26	2·0-4·0	0·5-1·5	8·5-10·5	1·2	0·5	1·0	1·0	0·2	0·1	0·2	–
LM27	1·5-2·5	0·3	6·0-8·0	0·8	0·2-0·6	0·3	1·0	0·2	0·1	0·2	–
LM28	1·3-1·8	0·8-1·5	17-20	0·7	0·6	0·8-1·5	0·2	0·1	0·1	0·2	Cr 0·6; Co 0·5
LM29	0·8-1·3	0·8-1·3	22-25	0·7	0·6	0·8-1·3	0·2	0·1	0·1	0·2	Cr 0·6; Co 0·5
LM30	4·0-5·0	0·4-0·7	16-18	1·1	0·3	0·1	0·2	0·1	0·1	0·2	–

a — single figures in this table are maxima. b — 0·2% in castings

109

BS 1490	Sand casting	Chill casting	Die casting	Fluidity	Resistance to hot-tearing	Pressure tightness
LM0	F	F	F	F	P	F
LM2	G*	G*	E	G	E	G
LM4	G	G	G	G	G	G
LM5	F	F	F*	F	F	P
LM6	E	E	G	E	E	E
LM9	G	E	G*	G	E	G
LM10	F	F	F*	F	G	P
LM12	F	G	U*	F	G	G
LM13	G	G	F*	G	E	F
LM16	G	G	G*	G	G	G
LM18	G	G	G*	G	E	E
LM20	E*	E	G	E	E	E
LM21	G	G	G*	G	G	G
LM22	G*	G	G*	G	G	G
LM24	F*	F*	E	G	G	G
LM25	G	E	G*	G	G	G
LM26	G	G	F*	G	G	F
LM27	G	E	G*	G	G	G
LM28	P	F	—	F	G	F
LM29	P	F	—	F	G	F
LM30	*	F	G	G	G	F

*Not normally used in this form.
E—Excellent G—Good F—Fair P—Poor U—Unsuitable

AND OTHER PROPERTIES

Machin-ability	Resistance to corrosion	Specific gravity	Coefficient of linear expansion[1] (C × 10^6)	Thermal conductivity[2] at 25°C (c.g.s. units)	Electrical conductivity[2] at 20°C (% IACS)	BS 1490
F	E	2·70	24	0·50	57	LM0
F	G	2·74	20	0·24	26	LM2
G	G	2·75	21	0·29	32	LM4
G	E	2·65	23	0·33	31	LM5
F	E	2·65	20	0·34	37	LM6
F	E	2·68	22	0·35	38	LM9
G	E	2·57	25	0·21	20	LM10
E	P	2·94	22	0·31	33	LM12
F	G	2·70	19	0·28	29	LM13
G	G	2·70	23	0·34	36	LM16
F	E	2·69	22	0·34	37	LM18
F	G	2·68	20	0·37	37	LM20
G	G	2·81	21	0·29	32	LM21
G	G	2·77	21	0·29	32	LM22
F	G	2·79	21	0·23	24	LM24
F	E	2·68	22	0·36	39	LM25
F	G	2·76	21	0·25	26	LM26
G	G	2·75	21	0·37	27	LM27
P	G	2·68	18	0·32	—	LM28
P	G	2·65	16	0·30	—	LM29
P	G	2·73	18	0·32	20	LM30

1. Relates to temperature range 20-100°C.
2. These values are approximate and will vary with the condition of the casting.

TYPICAL MECHANICAL PROPERTIES

Alloy	Sand cast			
BS 1490[1] 1970	0·2% proof stress (N/mm²)	Tensile stress (N/mm²)	Elongation (%)	Brinell hardness number
LM0-M	30	80	30	25
LM2-M	–	–	–	–
LM4-M	100	150	2	70
LM4-TF	250	280	1	105
LM5-M	90	170	5	60
LM6-M	70	170	8	55
LM9-M	–	–	–	–
LM9-TE	120	180	2	70
LM9-TF	220	250	–	100
LM10-TB	180	310	15	85
LM12-M	–	–	–	–
LM13-TE	–	–	–	–
LM13-TF	200	200	–	115
LM13-TF7	140	150	1	75
LM16-TB	130	210	3	80
LM16-TF	240	280	1	100
LM18-M	70	120	5	40
LM20-M	–	–	–	–
LM21-M	130	180	1	85
LM22-TB	–	–	–	–
LM24-M	–	–	–	–
LM25-M	90	140	2·5	60
LM25-TE	130	170	1·5	70
LM25-TB7	100	170	3	65
LM25-TF	220	250	1	105
LM26-TE	–	–	–	–
LM27-M	90	150	2	75
LM28-TE	–	–	–	–
LM28-TF	120	130	0·5	120
LM29-TE	120	130	0·3	120
LM29-TF	120	130	0·3	120
LM30-M	–	–	–	–
LM30-TS	–	–	–	–

1. Suffix letters indicate the condition of castings: M—as cast; TE—precipitation treated; TB—solution treated; TB7—solution treated and stabilized; TF—solution and precipitation treated; TF7—fully heat-treated and stabilized; TS—stress relieved.

0·2% proof stress (N/mm²)	Tensile strength (N/mm²)	Chill cast Elongation (%)	Brinell hardness number	Alloy BS 1490[1] 1970
30	80	40	25	LM0-M
90	180	2	80	LM2-M
100	200	3	80	LM4-M
250	310	3	110	LM-4TF
90	230	10	65	LM5-M
80	200	13	60	LM6-M
—	200	3	—	LM9-M
150	250	2·5	80	LM9-TE
280	310	1	110	LM9-TF
180	360	20	95	LM10-TB
150	180	—	95	LM12-M
—	220	1	105	LM13-TE
280	290	1	125	LM13-TF
200	210	1	75	LM13-TF7
140	250	6	85	LM16-TB
270	310	2	110	LM16-TF
80	150	6	50	LM18-M
80	220	/	60	LM20-M
130	200	2	90	LM21-M
120	260	9	75	LM22-TB
110	200	2	85	LM24-M
90	180	5	60	LM25-M
150	220	2	80	LM25-TE
100	230	8	65	LM25-TB7
240	310	3	105	LM25-TF
180	230	1	105	LM26-TE
100	180	3	80	LM27-M
170	190	0·5	120	LM28-TE
170	200	0·5	120	LM28-TF
170	210	0·3	120	LM29-TE
170	210	0·3	120	LM29-TF
150	150	—	110	LM30-M
160	160	—	110	LM30-TS

Note. The typical properties shown in this table are those of separately cast test bars and may not be obtained in all parts of a casting.

RELATED OR APPROXIMATELY

BS 1490	BS aerospace	USA AA	USA commercial	USA SAE
LM0-M	—	—	—	—
LM2-M	—	—	—	—
LM4-M	—	319·0	319	326
LM4-TF	—	319·0	319	326
LM5-M	—	514·0	213	320
LM6-M	L.33	—	—	—
LM9-M	—	A360·0	A360	309
LM9-TE	—	—	—	—
LM9-TF	—	—	—	—
LM10-TB	L.53	520·0	220	324
LM12-M	—	222·0	122	34
LM13-TE	—	A332·0	A132	321
LM13-TF	—	A332·0	A132	321
LM13-TF7	—	A332·0	A132	321
LM16-TB	—	355·0	355	322
LM16-TF	L.78	355·0	355	322
LM18-M	—	443·0	43	35
LM20-M	—	413·0	13	305
LM21-M	—	319·0	319	329
LM22-TB	—	—	—	—
LM24-M	—	A380·0	A380	306
LM25-M	—	356·0	356	323
LM25-TE	—	356·0	356	323
LM25-TB7	—	356·0	356	323
LM25-TF	—	356·0	356	323
LM26-TE	—	F332·0	F132	332
LM27-M	—	—	—	—
LM28-TE	—	—	—	—
LM28-TF	—	—	—	—
LM29-TE	—	—	—	—
LM29-TF	—	—	—	—
LM30-M	—	390·0	390	—
LM30-TS	—	390·0	390	—

Suffix letters indicate the condition of castings: M—as cast; TE—precipitation treated; TB—solution treated; TB7—solution treated and stabilized; TF—solution and precipitation treated; TF7—fully heat-treated and stabilized; TS—stress relieved.

SIMILAR SPECIFICATIONS

USA ASTM	Canada commercial	France A57-702/3	Germany DIN 1725/2	Germany Alloy No.
995A	100	A5	–	–
–	–	–	–	–
SC 64D	117	–	–	–
SC 64D	117	–	–	–
G 4A	B320	A-G3T	G·AlMg'	245
S 12C	6290	A·S13	G·AlSi12	230
SG 100A	161	A·S10G	G·AlSi10ma	230/10
–	161	A·S10G	G·AlSi10Mg	230/10
–	161	A·S10G	G·AlSi10Mg	230/10
G10A	350	–	G·AlMg10	–
CG 100A	–	A·U10G	–	–
SN 122A	162	A·12UN	–	230/Ni
SN 122A	162	A·S12UN	–	230/Ni
SN 122A	162	A·S12UN	–	230/Ni
SC 51A	125	–	–	234
SC 51A	125	–	–	234
S 5A	173	–	–	–
S 12A/B	–	A·S13 Y4	G·AlSi12(Cu)	331
SC 64C	–	–	G·AlSi6Cu4	225
–	C117	–	–	–
SC 84A	C143	–	G·AlSi8Cu3	333
SG 70A	135	–	G·AlSi7Mg	–
SG 70A	135	–	G·AlSi7Mg	–
SG 70A	135	–	G·AlSi7Mg	–
SG 70A	135	–	G·AlSi7Mg	–
SC 103A	–	–	–	–
–	–	–	–	–
–	–	–	–	–
–	–	–	–	–
–	–	–	–	–
–	–	–	–	–
–	–	–	–	–
–	–	–	–	–

ALUMINIUM CASTING ALLOYS

Chemical

BS aerospace	Cu	Mg	Si	Fe	Mn
L33	0·1	0·10	10·0-13·0	0·6	0·5
L35	3·5-4·5	1·2-1·7	0·6	0·6	0·6
L51	0·8-2·0	0·05-0·20	1·5-2·8	0·8-1·4	0·1
L52	1·3-3·0	0·5-1·7	0·6-2·0	0·8-1·4	0·1
L53	0·1	9·5-11·0	0·25	0·35	0·10
L78	1·0-1·5	0·4-0·6	4·5-5·5	0·6	0·5
L91	4·0-5·0	0·10	0·25	0·25	0·10
L92	4·0-5·0	0·10	0·25	0·25	0·10
L99	0·10	0·20-0·45	6·5-7·5	0·15[b]	0·10
DTD361B	4·0-5·0	0·10	0·25	0·25	0·10
DTD716A	0·1	0·3-0·8	3·5-6·0	0·6	0·5
DTD722A	0·1	0·3-0·8	3·5-6·0	0·6	0·5
DTD727A	0·1	0·3-0·8	3·5-6·0	0·6	0·5
DTD735A	0·1	0·3-0·8	3·5-6·0	0·6	0·5
DTD741A	3·5-4·5	1·2-2·5	0·5	0·5	0·1
DTD5008B	0·1	0·5-0·75	0·25	0·5	0·1
DTD5018	0·2	7·4-7·9	0·25	0·35	0·1-0·3
DTD5028	0·10	0·20-0·45	6·5-7·5	0·15[b]	0·10

(a) Single figures in this table are maxima. (b) 0·20 in castings

116

Composition (per cent)[a]

Ni	Zn	Pb	Sn	Ti	Others
0·1	0·1	0·1	0·05	—	—
1·8-2·3	0·1	0·05	0·05	0·25	Si + Fe 1·0
0·8-1·7	0·1	0·05	0·05	0·25	—
0·5-2·0	0·1	0·05	0·05	0·25	—
0·10	0·10	0·05	0·05	0·25	—
0·25	0·10	0·05	0·05	0·25	—
0·10	0·10	0·05	0·05	0·25	—
0·10	0·10	0·05	0·05	0·25	—
0·10	0·10	0·05	0·05	0·20	—
0·10	0·10	0·05	0·05	—	Ti + Nb 0·05-0·30
0·1	0·1	0·1	0·05	—	Ti + Nb 0·2
0·1	0·1	0·1	0·05	—	Ti + Nb 0·2
0·1	0·1	0·1	0·05	—	Ti + Nb 0·2
0·1	0·1	0·1	0·05	—	Ti + Nb 0·2
0·1	0·1	0·1	0·05	—	Co 0·5-1·0; Nb 0·05-0·3
0·1	4·8-5·7	0·1	0·05	0·25	Cr 0·4-0·6
0·10	0·9-1·4	0·05	0·05	0·25	—
0·10	0·10	0·05	0·05	0·20	—

ALUMINIUM CASTING ALLOYS – AEROSPACE SERIES

Related or approximately similar specifications

Aerospace	BS 1490	USA AA	USA commercial	USA SAE	USA ASTM	Canada commercial	France A57-702/3	Germany DIN 1725/2	Germany alloy no.
L33	LM6-M	–	–	–	S 12C	6290	A-S13	G-AlSi12	230
L35	LM14-WP*	242·0	142	39	CN 42A	218	A-U4NT	–	–
L51	LM23-P*	–	–	–	–	–	A-S2U	–	–
L52	LM15-WP*	–	–	–	–	–	–	–	–
L53	LM10-TB	520·0	220	324	G 10A	350	–	G-AlMg10	–
L73	LM16-TF	355·0	355	322	SC51A	125	–	–	234
L91	LM11-W*	–	–	–	–	226	A-U5GT	G-AlCu4Ti	220
L92	LM11-WP*	–	–	–	–	226	A-U5GT	G-AlCu4Ti	220
L99	–	A356·0	A356	336	SG 70B	B135	–	G-AlSi7Mg	–
DTD361B	LM11-WP*	–	–	–	–	226	A-U5GT	G-AlCu4Ti	220
DTD716A	LM8-M*	–	–	–	–	B116	A-S4G	G-AlSi5Mg	235
DTD722A	LM8-P*	–	–	–	–	B116	A-S4G	G-AlSi5Mg	235
DTD727A	LM8-W*	–	–	–	–	B116	A-S4G	G-AlSi5Mg	235
DTD735A	LM8-WP*	–	–	–	–	B116	A-S4G	G-AlSi5Mg	235
DTD741A	–	–	–	–	–	–	–	–	–
DTD5008B	–	D712·0	40E	310	ZG 61A	432	A-Z5G	–	–
DTD5018	–	–	–	–	–	–	–	–	–
DTD5028	–	A356·0	A356	336	SG 70B	B135	–	G-AlSi7Mg	–
DTD5028	–	A356·0	A356	336	SG 70B	B135	–	G-AlSi7Mg	–

ALUMINIUM CASTING ALLOYS – AEROSPACE SERIES

Aerospace	BS 1490	0.2% proof stress[c] (N/mm²)	Sand cast Tensile stress (N/mm²)	Elongation (%)	0.2% proof stress[c] (N/mm²)	Chill cast Tensile stress (N/mm²)	Elongation %
							Minimum mechanical properties
L33	LM6-M	60[c]	162[d]	5	80[c]	185[d]	7
L35	LM14-WP*	210	220	–	230	280	–
L51	LM23-P*	125	160	2	140	200	3
L52	LM15-WP*	245	280	–	295	325	–
L53	LM10-TB	170	280	8	170	310	12
L78	LM16-TF	220	250	–	250	300	–
L91	LM11-W*	165	220	7	165	265	13
L92	LM11-WP*	200	280	4	200	310	9
L99	–	185	230	2	200	280	5
DTD361B	LM11-WP*	310[c]	324[d]	–	360[c]	402[d]	4
DTD716A	LM8-M*	80[c]	124[d]	2	90	162[d]	3
DTD722A	LM8-P*	130[c]	147[d]	1	130[c]	185[d]	2
DTD727A	LM8-W*	100[c]	162[d]	2.5	100[c]	232[d]	5
DTD735A	LM8-WP*	220[c]	232[d]	–	220[c]	278[d]	2
DTD741A	–	250[c]	263[d]	–	260[c]	340[d]	–
DTD5008B	–	150[c]	216[d]	4	180[c]	232[d]	5
DTD5018	–	170[d]	278[d]	5	170[d]	309[d]	10
DTD5028	–	193[ad]	263[ad]	5[a]	201[d]	278[d]	5
DTD5028	–	178[bd]	224[bd]	3[b]	–	–	–

*Obsolete
[1] Newton per square millimetre (N/mm²) = 0.0645 tons-force per square inch (tonf/in²)
[2] Ton-force per square inch (tonf/in²) = 15.4443 newtons per square millimetre (N/mm²)
(a) test pieces cut from designated locations of casting; (b) test pieces cut from undesignated locations.
(c) For information only for LM alloys and also for others where indicated; (d) conversion from tons per sq. inch.

ALUMINIUM CASTING ALLOYS — AEROSPACE SERIES

BS Aerospace	Specific gravity	Casting characteristics			General properties	
		Sand casting	Chill casting	Die casting	Corrosion resistance	Machin-ability
L33	2·65	E	E	G	E	F
L35	2·82	F	G	U	F	G
L51	2·77	G	G	G*	G	G
L52	2·75	F	G	U	G	G
L53	2·57	F	F	F*	E	G
L78	2·70	G	G	F*	G	F
L91	2·80	F	P	U*	F	G
L92	2·80	F	P	U*	F	G
L99	2·68	G	E	F*	E	F
DTD361B	2·80	F	P	U*	F	G
DTD716A	2·68	G	G	G*	E	F
DTD722A	2·68	G	G	G*	E	F
DTD727A	2·68	G	G	G*	E	F
DTD735A	2·68	G	G	G*	E	F
DTD741A	2·80	F	G	—	F	G
DTD5008B	2·81	F	P	U*	E	G
DTD5018	2·64	F	F	—	E	G
DTD5028	2·68	G	E	—	E	F

*Not normally used in this form
E—Excellent; G—Good; F—Fair; P—Poor; U—Unsuitable

ALUMINIUM CASTING ALLOYS

	Red	Yellow	Blue	Brown	Black	Green	White
White							0
Green						12	11*
Black					5	30	10
Brown				23*	14*	29	28
Blue			1*	26	3*	27	—
Yellow		6	25	9	13	20	18
Red	4	8	24	22	16	21	2

*Obsolete.

STANDARD MELTING AND FLUXING
PROCEDURES FOR ALUMINIUM ALLOYS

Aluminium — Silicon — Copper Alloys
e.g. BS 1490, LM4, LM13, LM16, LM21, LM22, LM24 and LM27

Metal Treatment for Sand Castings

Melt initial charge under a cover of 250 g of COVERAL 11 per 50 kg melt. Continue charging until melt is complete and has acquired requisite degree of superheat. When metal temperature begins to fall slowly plunge 0·25% DEGASER 190 or 0·35% DEGASER 450 by weight per melt. Hold until bubbling ceases. These amounts may be increased if the charges are corroded or dirty giving heavy gas contamination in the metal. Increased amounts of DEGASER are best plunged in two stages consecutively.

For certain applications the above combined degassing/grain refining treatment may be superseded by separate treatments with DEGASER 200 or DEGASER 400 in conjunction with NUCLEANT 2 (0·25 by weight), a powerful grain refiner. Whichever method is used, finish off by adding a further 250 g per 50 kg of COVERAL 11 to the metal surface, rabble gently and skim off cleanly.

Recommended pouring temperatures are:—
      Light castings, up to 13 mm section   730°C.
      Medium castings, 13 — 38 mm section   710°C.
      Heavy castings, over 38 mm section   690°C.

Metal Treatment for Gravity and Pressure Diecastings

A full metal treatment as outlined above is not usually necessary for gravity and pressure diecastings. What is very necessary, however, is that the holding bale out furnaces are regularly drossed off in order to circumvent the possible formation of hard spot inclusions in the metal. Gravity diecasters will normally hold metal at average temperatures, i.e. around 700—740°C, and can therefore use COVERAL 11 in the amounts and manner already described for drossing off.

Additionally, some gravity diecastings subject to shrinkage unsoundness, etc., benefit greatly from the deliberate introduction of a controlled amount of gas into the metal. Ejection of this gas from solution in the form of widely dispersed pinholes helps to offset possible shrinkage unsoundness during solidification. The gas (hydrogen) may be added by plunging DYCASTAL tablets into the melt (0·15% by weight) or a pinch of DYCASTAL powder may be added to the ladle before baling out.

Pressure diecasters, because of the different requirements of the process, often hold metal in the bale-out furnaces at temperatures only just above the liquidus, i.e. around 610—640°C. At these low temperatures normal drossing fluxes work too slowly if at all, and it becomes necessary to use COVERAL 73 which was specially designed for the purpose. About 0·5% by weight is scattered on the metal surface and lightly rabbled until the reaction is complete. Skim the dry dross cleanly

Grain Refinement

The grain refinement of commercial die-casting alloy improves feeding characteristics. This treatment reduces the incidence of shrinkage unsoundness in heavy cast sections and thin sections show less tendency to form laps and misruns. NUCLEANT 2 tabletted grain refiner is recommended and this should be plunged into the melt at the rate of 0·25% by weight, prior to drossing off the melt with COVERAL 11.

ALUMINIUM SILICON ALLOYS

(a) The Medium Silicon Group — 3—6% Silicon

For example Alcan B116; Alcoa 356; the DTD alloys 716A, 722A, 727A, 735A, LM18, LM25, and LM8 (obsolete).

Modification of these alloys is not essential, but it often improves mechanical properties. Grain refinement can also assist response to heat treatment where this is carried out, and will prevent patchiness due to grain-size variation in articles subsequently anodised.

Add 250 g COVERAL 36A per 50 kg melt as an initial cover. When the charge is molten add a further 250 g, wait until the flux liquifies and then work thoroughly into the metal. Plunge 0·1—0·2% NUCLEANT 2 by weight before degassing with 0·25% by weight DEGASER 190 or 0·35% DEGASER 450. Finally, dross off with a small amount of COVERAL 11. The clean, oxide-free modified metal produced by this technique will exhibit enhanced fluidity and improved mass feeding characteristics.

Suggested pouring temperatures for sand castings are:—

Light castings, under 13 mm section	730°C.
Medium castings, 13—38 mm section	710°C.
Heavy castings over 38 mm section	690°C.

(b) The Eutectic Alloy Group — 10—13% Silicon

For example; Aerospace L33, LM6, LM20; also sometimes known as Alpax, Silumin, etc.

General Notes

The BS 1490 specification covering the aluminium—silicon alloys LM6 and LM20 calls for the castings to be supplied in the as-cast condition, i.e. not heat-treated. As such they are suitable for exposed marine castings, motor-car and lorry fittings, and engine parts (manifolds, casings, etc.), switch boxes and generally where corrosion resistance and ease of casting are essential.

The aluminium—silicon alloys generally have excellent casting properties and very good resistance to corrosion; the LM6 type is the best in this latter respect. Although of medium strength with relatively low hardness and elastic limit, they have excellent ductility. Because of their high silicon content, they have a tendency to drag during machining and cause rapid tool wear. Carbide-tipped tools with large rake angles, low cutting speeds and the use of lubricants and coolants, are recommended.

They can be sand cast, die cast — both gravity and pressure — and are very suitable for low pressure casting.

Test Bars

Test bars are generally dry-sand moulded in Mansfield or similar sand or in oil sand. The bar with the 70 mm diameter head is called for under the BS specifications but where there is some choice in the matter many foundrymen prefer a bar with a larger diameter head. The dimensions of the BSS mould and machined bar are shown on page 135.

Bars are generally poured at a temperature of about 680°C; in no circumstances should the metal be overheated. At the start of pouring, the mould should be held at an angle of about 45° from the vertical and then brought gradually to the upright position as the head fills. Pouring should be carried out steadily and over a period of not less than 10 seconds.

Heat Treatment

The alloys LM6 and LM20 are supplied in the non-heat-treated condition. Apart from stress relieving and for improving ductility, heat treatment is of little interest. It is worth noting, however, that the high silicon alloys, Alpax beta and Alpax gamma covered by BS specification 1490LM9 (TE and TF) are subject to a precipitation treatment and a precipitation/solution treatment respectively.

Casting Characteristics

Fluidity :	Excellent — can be cast into thinner and more intricate sections than any other aluminium alloy.
Pressure Tightness :	Very good — especially suitable for castings required to be pressure tight.

| Hot-tearing : | Sand and die-cast castings show complete freedom from hot-tearing. |
| Patternmaker's shrinkage : | 1·3%. |

Melting Considerations

As with most aluminium alloys, accurate temperature control is essential if best results are to be obtained. Gas pick-up increases with temperature of the melt and the temperature should, therefore, be kept as low as possible consistent with the requirements of correct fluxing treatment. As with both gas and oil-fired furnaces, the temperature continues to rise after heating is stopped, the fuel should be shut off before the required peak temperature is reached. An efficient degassing treatment should always be carried out if the metal is to be used for sand castings.

Modification

Although aluminium—silicon alloys were the first alloys of aluminium to be produced, it was not until 1921, when Dr. Pacz found that their mechanical properties could be considerably improved by the addition of sodium fluoride, that they assumed any industrial importance. The process of adding sodium fluorides or other alkali fluorides to the molten metal to improve the mechanical properties was termed "modification". Later several investigators found that modification could be achieved by the use of other agents, such as alkali metals and alkaline—earth metals. Of these modifying agents, the two most widely used have been sodium metal and alkali fluorides. The use of sodium metal in its usual commercially available condition is hazardous and the use of alkali fluorides is laborious. Both of these methods have now been superseded by the vacuum treated sodium (Navac) process.

In unmodified aluminium—silicon alloys, the aluminium—silicon constituent is increased from 11·7% in the unmodified alloy to Photomicrograph No. 1 (magnification x 50), page 133. Although modification does not actually refine the grain size of the metal, it does break up this needle-like structure within the grains and changes it to the fine dispersed form as shown in Photomicrograph No. 2 (magnification x50), page 134. At the same time the amount of silicon which can be combined in the aluminium—silicon constituent is increased from 11·7% in the unmodified alloy to about 13% in the fully modified state. Thus, modification can result in the removal of primary (i.e. uncombined) silicon from the alloy.

The effect of modification on all the mechanical properties of the alloy is very marked, but is possibly greatest on the ductility and toughness. The following table gives typical figures before and after modification.

125

Mechanical properties (Typical figures in Avoirdupois units)	Unmodified alloy		Modified alloy	
	Sand cast	Chill cast	Sand cast	Chill cast
Ultimate tensile strength (ton/in^2)				
Minimum	6·0	9·0	10·5	13·0
Average	7·0	10·0	11·5	14·5
Elongation (% on 2 in.)				
Minimum	1·0	2·0	5·0	7·0
Average	3·0	4·0	8·0	14·0
Impact resistance, Izod (ft/lb)				
Minimum	—	—	4·0	6·0
Average	0·5	1·0	5·0	7·0

Reversion

The modified structure of the aluminium—silicon alloy is unstable and it tends to change back or revert to the unmodified condition. This change is called "reversion" and the rate of change is directly proportional to silicon content, temperature of the melt and time of standing and indirectly proportional to size of melt. Reversion is slow at temperatures below 750°C and does not occur to any considerable extent with a melt of about 50 kg or so during a 10-min holding period provided it is left undisturbed. Larger melts of 75 kg or more can be left longer without harm, but about 20 minutes should be regarded as a maximum period for all sizes of melt.

Advantage of using Vacuum Treated Sodium (Navac)

Modification may be carried out either by the use of metallic sodium or sodium in the form of salts. Each method has advantages and disadvantages, but both are efficient ways of adding sodium to the alloy. The salts method is clean and safe to handle, but is laborious and the salts have a tendency to pick up moisture. Metallic sodium is cheap but hazardous to handle. To protect it from moisture in the atmosphere it is normally stored under paraffin. Sodium which has been stored in this manner still contains a quantity of paraffin which is a source of hydrogen, and consequently aluminium alloys treated with metallic sodium of this type are usually gassy and sand castings produced from them may be porous. A 1 oz. (28 g) cube of sodium can contain as much as 0·2 g of oil which could produce 350 cm^3 of hydrogen gas. Due to reaction with moisture metallic sodium which has been exposed to the air is coated with a thin film of sodium hydroxide, which will decompose when the sodium is plunged into the molten aluminium to form sodium oxide and water vapour, and the latter is a source of hydrogen which is readily absorbed by the metal. There is also a strong possibility that commercially produced sodium contains a quantity of hydrogen in solid solution.

The FOSECO organisation has developed a product which has all the advantages and none of the disadvantages of the two previous methods of modification already discussed; this is NAVAC. NAVAC is metallic sodium specially treated to eliminate completely all traces of hydrogen, paraffin and hydroxide, and is packed in air-tight capsules of pure aluminium to prevent contamination of sodium by moisture in the atmosphere. Packed in this way, sodium can be stored for unlimited periods of time, provided the containers are undamaged.

It is, therefore, strongly recommended that if completely gas free metal is required, as is the case in sand casting, modification should be carried out with the appropriate quantity of NAVAC, which is supplied in accurately weighed units in sealed aluminium containers.

Advantages of Degassing before Modifying

It is sometimes thought that by using modified ingot as supplied by the refiner, subsequent modification processes in the foundry can be cut down or even dispensed with. However, when it is borne in mind that the degree of modification in as-supplied ingot varies considerably and also that the ratio of scrap to ingot in any one charge also varies it will be seen that the only satisfactory method of securing a constant degree of modification is to destroy any existing modification and then remodify to a known and controlled extent.

The main function of degassing is, of course, to remove harmful gases, mainly hydrogen. Where completely gas-free metal is required as in sand casting particularly, a strong degassing treatment with DEGASER 190 should be carried out. However, degassing not only removes harmful gases, it removes sodium as well, and so destroys modification. *It is essential, therefore, that degassing should precede modification and never vice versa.*

Melting Procedure (NAVAC Process)
1. Heat up crucible.
2. Charge in ingot metal and scrap (runners, risers, etc.).

3. As soon as the first part of the charge reaches a pasty condition, sprinkle over its surface sufficient COVERAL 11 to form a cover. Usually about 250 g per 50 kg is required.

4. Add remainder of charge and melt rapidly, keeping the protective flux cover intact as much as possible.

5. Do not permit metal temperature to exceed about 760°C, then stop heating or remove the crucible from the furnace. There should be sufficient heat in the crucible to keep the temperature of the melt from falling, and during this period the melt should be degassed with DEGASER 190 or 301, or the ring-shaped tablet, DEGASER 450.

6. Push the surface flux cover a little to one side and through the cleaned area plunge to the bottom of the melt the DEGASER Tablets, using 0·25% by weight DEGASER 190 or 0·35% by weight DEGASER 450 and hold down until the bubbling action has ceased — usually 2 to 3 mins. DEGASER 190 is a solid tablet and is plunged by means of a perforated bell plunger. DEGASER 450 are pierced tablets and these are simply placed on a hooked steel rod plunger. Coat the plungers well with FIRIT.

7. With the metal temperature between 720° and 740°C (1328—1364°F), plunge a NAVAC unit to the bottom of the melt, using a perforated bell plunger. When the reaction has subsided the plunger is raised and lowered two or three times to stir the metal without breaking the surface, and it is then withdrawn. A NAVAC 25 unit is suitable for a melt of 50 kg. For larger or smaller melts see the table on the NAVAC Leaflet No. B9.

8. Sprinkle a further quantity of COVERAL 11 on the surface of the melt, once again using 250 g per 50 kg and allow the melt to stand quietly for about five to ten minutes. The standing time will depend on the size of the melt and should not exceed 5 minutes for melts of less than 50 kg.

9. The flux layer should now be rabbled into the immediate surface of the melt until a red glowing dross is obtained. Resulting from this exothermic action, a fine powdery dross exceptionally free from metal will be obtained. Skim cleanly.

10. Pour without delay.

N.B.—It is emphasised that hard and fast rules on modification technique cannot be laid down ; the above procedure is given only as a guide. It is only by trial and error and exercising a great deal of intelligence that optimum results will be obtained.

Melting Procedure (Salts Methods)

1. Heat up crucible.
2. Charge in ingot metal and scrap (runners, risers, etc.).
3. As soon as the first part of the charge reaches a pasty condition sprinkle over its surface sufficient COVERAL 11 to form a cover. Usually about 250 g per 50 kg is required.
4. Add remainder of charge and melt rapidly, keeping the protective flux cover intact as much as possible.
5. Do not permit metal temperature to exceed about 760°C then remove the crucible from the furnace. There should be sufficient heat in the crucible to keep the temperature of the melt rising. After the maximum temperature has been reached, the melt should be degassed and grain refined on a falling temperature with DEGASER 190 or the ring-shaped DEGASER 450.
6. Push the surface flux cover a little to one side and through the cleaned area plunge to the bottom of the melt the DEGASER Tablets, using 0·25% by weight DEGASER 190 or 0·35% by weight DEGASER· 450 and hold down until the bubbling action has ceased — usually 2 to 3 minutes. DEGASER 190 is a solid tablet and is plunged by means of a perforated bell plunger. DEGASER 450 are pierced tablets and these are simply placed on a hooked steel rod plunger. Coat the plungers well with FIRIT.
7. Rabble the surface flux cover into the surface of the melt and skim cleanly. When the melt has reached a temperature of at least 790°C — and not before — spread over the cleaned surface, a layer of COVERAL 29A, using 500 g per 50 kg.
   (Alternatively, COVERAL 36A can be used, in which case the temperature of the melt should be 750°C and the quantity used 1 kg per 50 kg.)
   When the flux layer has liquefied — and not before — it should be well worked into the melt for about 3 or 4 minutes, using a plunger in a stirring action.
8. Sprinkle a further quantity of COVERAL 11 on the surface of the melt, using 250 g per 50 kg and, depending on the size of the melt, allow to stand quietly for about 5 to 10 minutes — e.g. 5 minutes for a 25 kg and 10 minutes for a 50 kg melt — or until the correct pouring temperature is reached.
9. The flux layer should now be rabbled into the immediate surface of the melt until a red glowing dross is obtained. Resulting from this exothermic action, a fine powdery dross exceptionally free from metal will be obtained. Skim cleanly.
10. Pour without delay.

N.B.—It is emphasised that hard and fast rules on modification technique cannot be laid down ; the above procedure is given only as a guide. A certain degree of skill is required to ensure that the optimum results are obtained.

Grain Refinement

As the maximum mechanical properties of the aluminium—silicon alloys are developed through modification as distinct from grain refinement, the latter process is generally not considered of particular importance. Nevertheless, heavy section sand castings can benefit quite considerably if the metal is given a strong grain refining treatment in addition to that produced during the degassing process with DEGASER 190 or 450. This can be carried out using NUCLEANT 2 tablets at the rate of 125 g per 50 kg. The tablets are plunged either before or simultaneously with the degassing treatment.

Casting Temperatures

Casting temperatures vary with section thickness but generally fall within the following limits :—

Light castings :	under 13 mm section	730°C.
Medium castings :	13 to 38 mm section	710°C.
Heavy castings :	over 38 mm section	690°C.

Moulding Technique

Aluminium—silicon alloys can be cast satisfactorily in green or dry sand moulds, in either natural or synthetic moulding sand, the main requirements of which should be high permeability and fairly low green strength. Suitable properties of both types of sand are as follows:—

	Natural Sand	Synthetic Sand
Grain fineness—AFS No. ...	120	80—100
Clay content	10%	4%
		(bentonite or similar clay)
Moisture	6—8%	3—4%
Permeability—AFS No. ...	20—30	40—50
Green compression strength	7lb in^2	5lb in^2

Green sand moulds should be dusted with TERRACOTE 20 and sleeked to improve casting finish. Alternatively, an excellent finish will be obtained by applying MOLDCOTE 9. This spirit-based dressing is ignited and burns off from the treated surface to leave a firm, closely adhering coating and gives the effect of skin-drying or torching.

Dry sand moulds and oil sand cores should be dressed with a water suspension of TERRACOTE 2 for best quality work, or with TERRACOTE 20 for ordinary commercial work. Both moulds and cores should, of course, be thoroughly dried before casting.

Running and feeding methods should be as for a high shrinkage alloy, i.e., large diameter runners and large feeding heads are necessary. Chills may have to be used also on heavy sections.

Castings should be gated in such a way as to promote directional solidification, i.e., the sections furthest away from the gates and heads should solidify first. With complex castings, the use of multiple ingates is recommended as this method, in conjunction with as low a pouring temperature as is possible, tends to minimise localised overheating of the mould face and helps to promote rapid and even solidification throughout the casting. Runners and gates should be cut in such a way as to allow the metal to enter the mould fairly rapidly but at the same time with minimum turbulence so as to avoid the entrapment of air and the formation of oxides.

Feeder heads must be of such size as to ensure that the feed metal remains liquid and available to the casting until the casting itself has solidified. If plain sand heads are used this means that the risers must often be of substantial size thus decreasing yield very considerably. Efficiency can be significantly improved by lining the riser cavity with a highly insulating resilient refractory sleeve of KALMIN 33. These are available in a range of ready made sizes and shapes but may be cut and shaped easily with a sharp knife to fit an unusual location. All riser heads should be covered with a layer of FEEDOL 9 immediately they are full in order to prevent heat loss to atmosphere.

Gravity Diecasting

In view of the effect of shock cooling on the metal it is sometimes considered unnecessary to modify aluminium—silicon alloy which has to be gravity diecast. Generally, however, metal for casting sections 5 mm and over should be modified for best results.

Since, in gravity diecasting, a small amount of gas in the metal is beneficial in countering shrinkage, modification can be carried out using NAMETAL. The use also of DYCASTAL which induces limited and controlled gassing of the melt, should be considered. Neither of these recommendations apply, however, to castings which have to pass radiological examination.

NAMETAL is used in the same way as NAVAC but at the rate of

131

about 45 g per 50 kg and plunged at a temperature of 750°C. Further regular additions of about 30 g per 50 kg of melt should be made every half or one hour depending on rate at which more metal is added. (Care must be taken nòt to add too much sodium metal as excess tends to make the metal sluggish and die dressings will be blackened.)

Pressure Die Casting

It is generally agreed that aluminium—silicon alloy for pressure die casting does not require to be modified. The temperature of the metal as injected is very low and this, coupled with shock cooling, combine to produce a spontaneous modification. Nevertheless, where there are any somewhat heavy sections the metal undoubtedly benefits from a mild modification. In such cases 20 g of NAMETAL per 50 kg of metal should be used.

75C

Photomicrograph No. 1
x 50 magnifications.
Showing
unmodified structure.

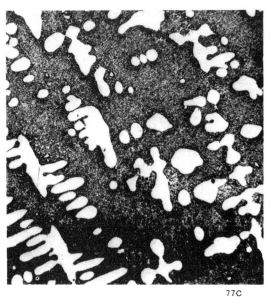

77C

Photomicrograph No. 2
x 50 magnifications.
Showing
modified structure.

STANDARD TEST BAR FOR
ALUMINIUM ALLOYS SAND CAST
Tensile Test Piece

The ends to be shaped to fit the axial loading shackles of the testing machine.

Cross-sectional area A_0	Diameter D	Gauge length L_0	Minimum parallel length P	Minimum radius at shoulder R
mm^2	mm	mm	mm	mm
150	13.82	69	76	26

Proportional Test Piece

All dimensions are nominal and are in mm.

ALUMINIUM-MAGNESIUM CASTING ALLOY
BS 1490 — LM5

The melting, fluxing and degassing technique described for LM5 applies equally to LM14 and also to the L.52 alloy.

Related Specifications are:
ALCAN GB. B320   ALCOA 214

Test Bars

Sand-cast test bars should conform to the drawings given on page 136. The dimensions given are those stipulated in BS 1490.

Test bars are generally moulded in dry sand or oil sand, as the specification stipulates a dried mould: this eliminates the possibility of variable chilling due to a variable moisture content. The steel tube for the mould should be perforated so as to form vents for the escape of mould gas. It should be not less than 80 mm in diameter.

The bars should be poured at a temperature of 690–700°C (1274–1292°F) and it is important if maximum physical properties are to be obtained that the mould be at atmospheric temperature. In no circumstances should the alloy be overheated. At the start of pouring, the mould is held at an angle of about 45° from the vertical and then brought gradually to the upright position as the head fills. Pouring is carried out steadily and should occupy a period of not less than ten seconds: turbulence within the test bar mould should be avoided at all costs.

Heat Treatment

None is required.   The BS 1490 Specification calls for the alloy LM5 to be supplied in the as-cast (M) condition.

Physical Properties

Density	2·66
Electrical conductivity (% IACS at 20°C)	29 approx.
Brinell Hardness (sand cast)	60.

Founding Characteristics

Fluidity	Fair.
Hot tearing	Fair resistance.
Pressure tightness	Poor.
Machinability	Good.
Resistance to corrosion	Excellent.

Melting Hazards (Gas absorption)

Aluminium—magnesium alloys of the LM5 type tend to oxidise and pick up hydrogen very readily during melting. Thus every precaution should be taken to protect the melt and the use of a low melting point flux with a strong cleansing action should be employed. The flux must also have the property of dissolving and coalescing oxides, particularly magnesium oxide. Such a flux is COVERAL 65.

As LM5 is susceptible to grain growth, a grain refining treatment is necessary. This is combined in the degassing action of DEGASER 190, which has the additional advantage that it forms magnesium chloride and so is beneficial in helping to free the melt from entrained oxides and non-metallics.

Melting Procedure

1. Heat up the crucible.
2. Charge ingot and scrap (runners, risers, etc.).
3. Dust the solid metal with a generous quantity of COVERAL 65, using about 250 g per 50 kg of melt.
4. Add the remainder of the charge and melt rapidly, keeping the fluid protective cover intact as much as possible.
5. When the melt has reached the plastic condition, i.e. at a temperature of about $600°C$ ($1110°F$), add a further quantity of COVERAL 65, spreading it evenly over the surface of the melt. A quantity of at least 1 kg per 50 kg is required.
6. On attaining a temperature of about $700°C$ ($1290°F$), stop heating or remove the crucible from the furnace. There should be sufficient residual heat in the refractory crucible and furnace lining to bring the temperature of the charge up to, but not exceeding, $750°C$ ($1380°F$). Careful control is necessary at this stage, as the temperature must not be allowed to exceed $750°C$.
7. The flux cover, which will now be in a fluid condition, is thoroughly "washed" into the molten alloy, using a perforated plunger in a stirring action, alternatively, a skimmer can be used, stirring in such a way as to bring the fluid flux into contact as much as possible with the molten metal.
8. Push the slag on the surface of the charge, a little to one side and plunge to the bottom, through the cleaned area. DEGASER tablets 190. Use at least 250 g per 50 kg. In the case of large melts, it is recommended that the DEGASER be plunged in at least two stages, using a proportion of the total quantity necessary at each stage. Hold the DEGASER tablets at the bottom of the melt until the bubbling action has ceased.

9. Sprinkle on the surface of the melt a quantity of COVERAL 11, using about 250 g per 50 kg and allow to stand undisturbed for about 5 minutes.

10. The flux is now rabbled into the *immediate surface only* until a glowing red dross is formed. As a result of this exothermic reaction, a fine powdery dross, exceptionally free from entrained metal, is obtained on skimming. Check the temperature and pour without any further delay.

To prevent iron pick-up, plungers and all other iron furnace tools are kept coated with a refractory wash of FIRIT. They must be dried and thoroughly heated before use.

The risk of iron pick-up is minimised and degassing activity enhanced by using DEGASER 450 as an alternative. These tablets are provided with a central hole so that plunging can be carried out by means of a simple hooked rod, eliminating bell plungers which trap and release the scavenging gas as large bubbles.

DEGASER 450 is plunged at the rate of 200 g per 50 kg of metal.

Casting Temperatures

These vary according to thickness of casting section, but generally fall within the following limits :

Light castings—
    under 13 mm section   ...   ... 720°C (1330°F)

Medium castings—
    13 to 38 mm section   ...   ... 700°C (1290°F)

Heavy castings—
    over 38 mm section   ...   ... 680°C (1260°F)

Sand Properties

Both naturally bonded and synthetic sands may be used, as well as silicate bonded sand and the sands containing synthetic resins as used for shell moulding. In Great Britain natural quarried sands such as Bromsgrove and Mansfield red, are quite suitable. Typical properties are indicated in the following table:—

	Natural sand	Synthetic sand
Grain fineness—AFS No.	120	80–100
Clay content (%)	10	4
Moisture (%, green) ...	6	4·5–5·0
Permeability—AFS No.	50	100–120
Green compression strength (lbs. per in^2) ...	7–8	5–6

139

Metal—Mould Reaction

Although metal—mould reaction is greatest in magnesium bearing alloys of the LM10 type, which contain 10% of magnesium, it is also very pronounced in the LM5 alloys. Its occurrence causes the casting surface to show patches of oxidation which may be severe and deep seated in thick sections. Increase in magnesium content and the length of time taken for the casting to solidify magnifies the trouble, which is not avoided by the use of dried sand moulds and cores.

This metal—mould reaction results from the action of the magnesium present in splitting up the moisture in the sand into its constituents hydrogen and oxygen. The latter forms oxide with the alloy and the hydrogen, in the atomic state, is released into the molten metal. As the casting solidifies the hydrogen is expelled from solution and porosity is the result. This porosity is concentrated in the outer layers of the casting section but diminishes towards its centre. Even with dried sand moulds and cores there is sufficient water combined with the clay bond to react with the magnesium in this way.

J. Lund has shown that mould reaction in the aluminium 5% magnesium alloys is reduced by hard ramming and by increasing the moisture content of green sands. This, in effect, means that influences which increase the rate of freezing of the casting, lessen the severity of the mould reaction. It follows, therefore, that increase of pouring temperature will cause a higher degree of porosity.

A particularly efficient mould or core dressing designed to reduce and often eliminate metal—mould reaction is MOLDCOTE 41, a self-drying type of dressing which contains inhibitors.

Facing Sand Inhibitors

Metal—mould reaction can largely be prevented in the aluminium alloys containing magnesium, by making additions of chemicals known as inhibitors to the facing sand. The inhibitor most used is boric acid, and it is added at the rate of about 2%, although more may be needed in the case of very thick sectioned castings. Alternatively chills may be employed in such cases to increase the rate of cooling.

Ammonium bifluoride, added to the facing sand in the same ratio, may be used in place of boric acid. However, the latter is usually preferred because it does not produce objectionable fumes when the metal is poured but it has the disadvantage that it tends to adhere to the casting and causes a hardening of the sand which hinders knocking-out.

Silicate Bonded Sand

Neither boric acid nor ammonium bifluoride should be added to sand used in the carbon-dioxide process as they exert a deleterious effect. Fortunately, when employing silicated bonded mould and cores, the use of these inhibitors appears to be unnecessary provided that a suitable dressing or coating is used.

Shell Moulding

If urea formaldehyde resins are used, this gives a natural inhibiting effect. Boric acid can be used with phenol formaldehyde resin and has an additional advantage in helping the curing process.

Segregation of Inhibited Sand

It is most important to note that facing sands containing an inhibitor such as boric acid, give rise to pronounced metal—mould reaction if employed in moulds for modified aluminium—silicon alloys. Such sand must be kept away from that used for these alloys and for those not containing magnesium, especially the high silicon alloys like LM6.

Mould Dressings

Green sand moulds should be dusted with TERRACOTE 20 and sleeked to improve coating finish. Alternatively an excellent finish will be obtained by applying MOLDCOTE 6 or 9. These spirit based dressings are ignited and burn off from the treated surface to leave a firm, closely adhering coating, and give the effect of skin drying or torching.

Dry sand moulds and oil sand cores should be treated with a water suspension of TERRACOTE 2 for best quality work, or with TERRACOTE 20 for ordinary commercial work. Both moulds and cores must be thoroughly dried after applying these dressings.

Running and Gating Methods

Bearing in mind that aluminium-magnesium alloys tend to oxidise easily and form heavy drosses, runners and gates should be formed in such a way as to allow the metal to enter the mould fairly rapidly but with a minimum of turbulence. This will help to avoid entrapment of air and the formation of oxides. For large castings a ring runner with multiple ingates gives good results. Bottom pouring or gating near the base of the mould is the general rule. Also perforated, thin steel plates, in the runner systems assist the production of castings free from inclusions. They help, too, in choking the downgate and so are effective in keeping up the head of metal in the runner bush which guards against dross being carried into the mould.

Feeding Methods

Generous feeding heads are necessary. On heavy sections and when casting design prevents the placing of adequate feeding heads, denseners or chills should be used. They must be clean and preferably coated with a dressing such as CHILCOTE 7 and dried thoroughly before use.

Although chills and denseners are most often made of cast iron or steel, there is something to be said for producing them in aluminium alloy so far as light alloy founding is concerned. One obvious advantage is that they can be cast from the molten metal handled in the shop. Other advantages are their high heat conductivity, specific heat and latent heat of fusion. Also they are easier to machine or file and, being light, are much easier to hold in position in a green sand core or mould. For best results, the face of the chill should be grooved to allow for escape of air at the interface. A disadvantage of the use of light alloy chills, particularly in foundries with a mechanised sand system, is that they do not respond to magnetic separation from the knock-out sand.

As with all castings some thought should be given to obtaining as much directional solidification as possible. Those parts of the casting furthest from the ingates should solidify first and feeding should be progressive back to the risers. The feeder heads must remain molten longer than the casting or feeding cannot take place. It is important that risers are not over-large so reducing yield and also important that heat loss from the riser mass is retarded as far as possible.

For this purpose, feeder sleeves for lining the riser cavity can be employed. These sleeves can be obtained in FEEDEX exothermic material or the newer and more effective lightweight, highly insulating KALMIN 33 shapes can be used. Insulating sleeves have low specific heat, extremely good long-term properties, and are completely inert to all aluminium alloys. (They may also be used satisfactorily on magnesium base alloys.)

Whichever form of open-top riser is used, it is important to reduce heat loss to atmosphere as far as possible. This is increasingly important with KALMIN 33 sleeves where loss to atmosphere is the major route for heat loss. All risers should therefore be covered with a layer of FEEDOL 9 anti-piping compound immediately on completion of pouring. The layer should be about 15 mm thick or as thick as the riser sleeve material. Using these techniques, it should be possible to eliminate rod feeding, with all its objections, completely.

Gravity Die Casting

Dies should be designed with the runner-riser system cut on generous lines similar to the practice used for LM14 (Y alloy) and with a view to creating the minimum of turbulence as the stream of metal enters. In some cases slightly thicker sections than usual may be required to obtain the best results in castings.

Melt down the ingot and process scrap under a layer of COVERAL 65. Superheat to a maximum temperature of 750°C and skim. Add a further 2% of the charge weight of COVERAL 65. Allow the flux to fuse on the surface, followed by a thorough puddling of the "wet" slag for 3—4 minutes, employing for the purpose a skimmer tool, making a rotary motion. Any remaining "islands" of slag or dross should be skimmed off and the sides of the crucible, at the metal level, gently scraped. Ladling out then proceeds in the usual manner.

In the case of a bale-out furnace working continuously with more ingot metal and scrap being fed from time to time, the routine outlined is repeated at intervals dependent upon the rate at which the metal is consumed.

These alloys are susceptible to grain growth, even when cooled relatively quickly as in gravity die casting. To ensure the production of a fine grain size, NUCLEANT 2 tablets are plunged on a basis of 0·25% by weight.

Every endeavour should be made via die design, casting technique, etc., to obtain sound castings using the standard metal treatment outlined.

Since most LM5 castings are radiographed, anodised, or surface finished in some way, it is normally undesirable that recourse is made to introducing controlled amounts of gas to the melt with DYCASTAL.

An appropriate DYCOTE insulating die dressing should be applied to the die to control solidification and obtain optimum surface finish.

ALUMINIUM—MAGNESIUM CASTING ALLOY
BS 1490 — LM10

BSS (Aerospace) L53.
BS/STA7 : AC10.   Hiduminium 90.
American ASTM ; B179-49T-G10A.
American ALCOA 220.   Canadian ALCAN GB350.
French A-G11.
German DIN 1725.   AIMg7 and GD AIMg 9.

Experience shows that the magnesium content should be maintained at the higher level of the specification as this helps in obtaining the specified mechanical properties. However, it should not exceed 11% in the interests of maximum resistance to corrosion fatigue. Small quantities of beryllium exert a marked effect in reducing metal/mould reaction. Further comments on this point are included later.

Casting Tensile Test Bars

Tensile test pieces of LM10 alloy should be cast to shape in green sand using a multiple test bar mould known as E Bar pattern. The drawing and dimensions of this mould pattern are shown at the end of these notes.   The bars should be cut from the running system before placing in the heat treatment furnace : this will prevent distortion.   Bars should not be heat treated in the machined state.

Heat Treatment

Specification BS 1490 calls for castings in this alloy to be supplied in the solution heat-treated (TB) condition. It consists of heating for a minimum period of 8 hours at a temperature between 425° and 435°C (797° and 815°F). Eight hours is the minimum holding time for average castings; it may have to be increased in special cases. Following this the castings may be allowed to cool in the furnace to 385° to 395° before quenching which consists of immersion in oil, the oil having a temperature not in excess of 160°C (320°F). The castings may remain in the oil no longer than one hour and are finally quenched in water or cooled in air to atmospheric temperature.

Physical Properties

Specific gravity	...2·55 (sand cast)	2·60 (die cast)
Weight per cubic inch (lb)	0·092
Specific resistance, microhms/cm^3	8·5
Electrical conductivity (%IACS at 20°C) ' ...	20 approx.

Founding Characteristics

Fluidity	Can be described as fair only and relatively large diameter runners and adequate ingates are required. It is not very suitable for thin sectioned castings.
Pressure tightness ...	Poor : it is difficult to make castings which have to be pressure tight.
Hot tearing	Resistance to hot tearing is good.
Patternmakers' shrinkage	1·3%

Machinability also is good.

Melting Hazards (Gas absorption)

Aluminium— magnesium alloys of the LM10 type tend to oxidise and pick up hydrogen very readily during melting. Thus every precaution must be taken to protect the melt and the use of a flux with a strong cleansing action and able to dissolve and coalesce oxides, particularly magnesium oxide, is essential. Such a flux is COVERAL 33FF which has the further advantage of being completely free from sodium. Contamination of the melt with sodium must be avoided because its presence is detrimental to the development of full mechanical properties.

A further factor which influences the development of maximum mechanical properties is the grain size which should not exceed 0·10 to 0·15 mm diameter. Immediately before degassing the melt should be grain refined by making simultaneously small additions of titanium and boron such as are contained in NUCLEANT 2 tablets.

Aluminium—magnesium alloys of LM10 composition are prone to patches of porosity rather similar to those which form in magnesium base alloys, in which it is referred to as "lake" porosity. Porosity of this kind is very harmful to mechanical properties and the strength of test bars so affected may drop by as much as 70%. It is therefore extremely important that a correct metal treatment is strictly adhered to and also that a suitable test bar mould and casting technique is used.

The melt must not be overheated, and a temperature of $720°C$ ($1330°F$) should be kept as a maximum. In this way gas pick-up is reduced and large grain growth suppressed. Accurate temperature control is essential to success.

Melting Procedure

1. Heat up crucible and place in the bottom a quantity of COVERAL 33FF (in the form of "chippings") using 500 g per 50 kg. Where tall, narrow crucibles are used, the quantity may be less as the 1% ratio applies to melting in bale-out and reverberatory furnaces of conventional surface area.

2. Charge ingot metal and scrap (runners, risers, etc.).

3. Melt in the usual way, adding more ingot or scrap until the charge is complete.

4. When the melt has just passed the "plastic" stage add a further quantity of COVERAL 33FF, using about 1·5 kg per 50 kg, spreading it evenly over the metal surface.

5. Continue heating until the temperature reaches 680°C (1250°F), then shut off the fuel supply or remove the crucible from the furnace.

6. The residual heat in the crucible or refractories will raise the temperature to about 720°C and the layer of COVERAL 33FF will be in a semi-liquid condition so that it can be well rabbled into the molten charge for from three to four minutes. Use a plunger and stir so as to bring the flux into contact with the molten metal as much as possible.

7. Now grain refine by plunging NUCLEANT 2 tablets to the bottom of the molten charge. Use 125 g per 50 kg and hold submerged until all bubbling ceases.

8. Follow by degassing with DEGASER 200 tablets. These also are plunged as for NUCLEANT but use 250 g per 50 kg. If the melt is large, degassing is best carried out in two or more stages using appropriate proportions of the total quantity of tablets at each stage. After degassing do not interfere with the melt until it is poured.

Note.—The quantity of COVERAL 33FF mentioned refers to melting in bale-out and reverberatory furnaces. Where tall narrow crucibles are used, the amount may be reduced to some extent. NUCLEANT and DEGASER tablets must be plunged to the bottom of the melt. This can be done most easily and conveniently by means of a special perforated plunger. This tool is also useful for rabbling in the COVERAL 33FF flux. To restrict iron pick-up, plungers and all other iron furnace tools should be coated with a refractory wash of FIRIT and dried thoroughly before use.

Casting Temperatures

These vary with section thickness but fall, generally, within the following limits :

Light castings—
under 13 mm section ... ... 720°C (1330°F)

Medium castings—
13—38 mm section ... 700°C (1290°F)

Heavy castings—
over 38 mm section ... ... 680°C (1260°F)

Metal—Mould Reaction

It has already been pointed out that the inclusion of 0·004% beryllium in the LM10 alloy helps to reduce metal/mould reaction. None of the common impurities, except sodium, interferes with this beneficial action of beryllium.

When metal—mould reaction occurs the surface of the casting shows oxidation which may be severe and deep seated. Hydrogen, too, is absorbed and liberated, as the casting solidifies, to cause porosity. This porosity is concentrated beneath the casting surface, but diminishes towards the centre of the section. It results from the splitting up of the moisture in the sand due to the presence of magnesium. The oxygen forms oxide with the alloy and the hydrogen is released into the molten metal. Increase in magnesium content and length of time taken for the casting to solidify magnifies the trouble, which is not avoided by the use of dried sand moulds and cores.

Apart from the use of beryllium, which is effective in overcoming metal—mould reaction in the thinner sectioned range of castings, inhibiting agents are added to the facing sand. Thus for casting sections above 25 mm or about 13 mm plate thickness, 2% of boric acid or a similar quantity of ammonium bifluoride is milled into the facing sand. Boric acid is usually preferred because it does not produce objectionable fumes when the metal is poured. However, sand containing boric acid, tends to adhere to the casting and the hardening of the sand hinders knocking-out. For very thick sections a higher proportion of inhibitor will be necessary.

M. Whitaker has worked on the problem of metal—mould reaction in the 10% magnesium—aluminium alloy and found that it is influenced by mould variables such as ramming density, moisture content, and coarseness of the sand grains. These factors affect the chilling power of the mould which should be at a maximum to reduce the reaction. The soundest castings are obtained by using medium or fine sand and ramming as hard as is consistent with adequate permeability and other desirable characteristics.

Other important factors are the maintenance of 0.004% beryllium in the alloy, the use of inhibiting agents in the sand and pouring at a temperature as low as possible consistent with good foundry practice.

Silicate Bonded Sand

Neither boric acid nor ammonium bifluoride should be added to sand used in the carbon-dioxide process as they have a deleterious effect. Fortunately, when employing silicate bonded moulds and cores, the use of these inhibitors appears to be unnecessary. In certain cases, particularly for heavy cast sections, a dusting of INERTEX over mould and core surfaces is advised as a safeguard.

Shell Moulding

If urea formaldehyde resin is used, this gives a natural inhibiting effect. Boric acid can be used with phenol formaldehyde resin and has an additional advantage in helping the "curing" process.

Loss of Beryllium on Re-melting

Some loss of this element does occur during remelting the alloy. Therefore, an occasional check should be made to ensure that the optimum content of 0.004% is maintained. The loss is greatest in small melts, and it is suggested that for melts of about 50 kg a check should be made after about eight or so consecutive remeltings. Experience will determine whether more or less frequent checks are necessary.

Mould Dressings

Green sand, dry sand or CO_2 moulds and cores may be coated with MOLDCOTE 41, a self drying type of dressing, to diminish metal—mould reaction. This dressing may be applied by brush, swab or spray gun and takes approximately 10 minutes to air dry.

Running and Gating Methods

Bearing in mind that aluminium-magnesium alloys tend to oxidise easily and form heavy drosses, runners and gates should be formed in such a way as to allow the metal to enter the mould fairly rapidly but with a minimum of turbulence. This will help to avoid entrapment of air and the formation of oxides. For large castings a ring runner with multiple ingates give good results. Bottom pouring or gating near the base of the mould is the general rule. Also perforated, thin steel plates, in the runner system, assist the production of castings free from inclusions. They help too in choking the downgate and so are effective in keeping up the head of metal in the runner bush which guards against dross being carried into the mould.

Additional points to be observed are :

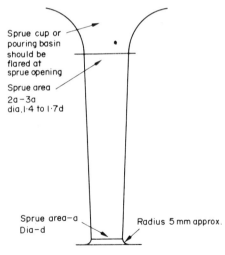

Sprue cup or pouring basin should be flared at sprue opening

Sprue area 2a – 3a dia. 1·4 to 1·7d

Sprue area—a Dia—d

Radius 5 mm approx.

1. Sprues or down runners should be tapered, and the smallest sprue that will prevent mis-runs used.

2. Sprue base should be enlarged or a well provided to slow down flow of metal entering the runner bar.

3. Runner bar should be cut in drag and the gates in the cope to ensure keeping runners full and to provide skimming action.

4. Total cross sectional area (CSA) of runner bar should be four to six times restricted CSA of Sprue. This ratio slows metal at gates.

5. Total CSA of gates should equal CSA of runner bar.

Recommended design for Sprue or down runner pattern for Sand Cast LM10.

6. CSA of runner bar should be reduced by CSA of gate as each gate is passed. This helps to keep runner bar and gates full and equalises the flow through each gate.

7. Abrupt changes in area of flow channel and in direction should be avoided as should sharp corners.

8. Castings should be fed through risers. For deep castings use slot gates entering casting from side risers.

When parallel-sided down runners are used there is a reduced pressure and gases are aspirated into the metal stream during pouring. The tapered runner helps to set up a back pressure and air and gases are not sucked in.

Gravity Die Casting

Dies should be designed with the runner-riser system cut on generous lines similar to the practice used for LM14 (Y alloy). In some cases slightly thicker sections than usual may be required to obtain the best results in castings.

Feeding Methods — Sand Castings

Feeder heads of generous proportions are essential. On heavy sections and when casting design prevents the provision of adequate feeding heads, denseners or chills should be used.

They must be clean and preferably coated with a dressing, such as CHILCOTE 7, and dried thoroughly before use.

Satisfactory chills for light alloy castings can be formed from cast iron, steel, brass or gunmetal. However, in a light alloy foundry chills can be made most readily from the alloys being cast. Their advantages are high heat conductivity, specific heat and latent heat of fusion. Also they are easier to machine or file ; and being light, are much easier to hold in position in a green sand mould or core. For best results, the face of the chills should be grooved to allow air to escape the metal mould interface. A disadvantage of the use of light alloy chills, particularly in continuous production foundries, is that they do not respond to magnetic separation from the knock-out sand. As in all cases, every effort should be made to promote directional solidification ; that is, ensure that both areas remote from ingates and feeder heads solidify first. Feeder head dimensions should be sufficient to delay solidification until the casting has practically set. Slightly tapering risers, wide end up, are advised ; the height to diameter ratio of risers being about 2 to 1.

To improve feeding efficiency, and at the same time reduce riser dimensions, exothermic or insulating sleeves can be of great benefit.

Prefabricated sleeves and shapes can be obtained in either FEEDEX 4 or KALMIN 33. Technically the highly insulating, inert KALMIN 33 has advantages over the exothermic sleeve and for these very reactive alloys increasing use is being made of them. Heat loss to atmosphere from the open top of the riser should be controlled by applying a layer of FEEDOL 9 hot topping compound immediately on completion of casting.

Segregation of Special Sands

Sand which contains any of the inhibiting agents described must be kept apart from other sand mixtures. For example, if inhibiting sand is used for moulds in which silicon bearing alloys are cast, particularly high silicon alloys such as LM6 (Silumin), pronounced metal—mould reaction can cause severe porosity. Likewise, sand containing sulphur, such as is used for magnesium-based alloys, should not be employed for moulds into which LM10 alloy is cast. Even as little as 0·5% sulphur is sufficient to cause metal—mould reaction, despite the presence of beryllium in the metal, or inhibitors in the facing sand.

These precautions are not necessary for silicate-bonded sands whether the hardening agent is carbon dioxide gas or one of the new chemical self-hardening systems.

Tensile test sample

Section A–A

$1\frac{23}{32}$" dia $1\frac{9}{16}$" dia

$2\frac{3}{4}$" $3\frac{17}{64}$" $1\frac{11}{16}$"

$2\frac{11}{16}$" 45° $\frac{1}{8}$" $2\frac{11}{32}$" 45° $\frac{1}{8}$" dia

$9\frac{1}{16}$" $9\frac{1}{16}$" $2\frac{3}{4}$" $2\frac{1}{8}$" $1\frac{7}{8}$" $2\frac{1}{16}$" $1\frac{9}{16}$"

0·625" dia 1·0" dia

View B–B

Section C–C $1\frac{11}{16}$" $1\frac{11}{16}$" $1\frac{25}{32}$" $\frac{63}{64}$"

Tolerances
Diameters ± 0·01
Other dimensions ± 0·02

$9\frac{1}{16}$ in.	230·18 mm	$2\frac{1}{16}$ in.	52·38 mm	1 in.	25·40 mm
7 in.	177·80 mm	$1\frac{7}{8}$ in.	47·62 mm	$\frac{63}{64}$ in.	25·00 mm
$3\frac{17}{64}$ in.	82·95 mm	$1\frac{25}{32}$ in.	45·25 mm	$\frac{11}{16}$ in.	17·46 mm
$2\frac{3}{4}$ in.	69·85 mm	$1\frac{23}{32}$ in.	43·65 mm	0·625 in.	15·88 mm
$2\frac{11}{16}$ in.	59·55 mm	$1\frac{9}{16}$ in.	39·69 mm	$\frac{1}{8}$ in.	3·17 mm
$2\frac{1}{8}$ in.	53·97 mm	$1\frac{7}{16}$ in.	36·51 mm		

152

ALUMINIUM-ZINC-MAGNESIUM CASTING ALLOY

FRONTIER 40E ALLOY*

Frontier 40E alloy is an aluminium-zinc-magnesium casting alloy, possessing high strength, pressure tightness and hardness obtained without resorting to heat treatment. It has excellent machinability, exceptional resistance to shock and corrosion, very good anodising properties and weldability. From the point of view of resistance to atmospheric and marine corrosion, it can be compared favourably with the aluminium magnesium alloys.

Related specifications are :

DTD.5008B (British).
612A and 612B Alcoa.
Tenzaloy, American Smelting
    and Refining Co.

Ternalloy 8, Apex Smelting
Unifont, AIAG.
Inafond, Z5F, Montecatini.
French Alliage A/Z 5G.
USA ASTM B-26-52T.

Chemical Composition of DTD.5008B Alloy

Zinc ...	...	...	4·8—5·7%
Magnesium	...	...	0·5—0·75%
Chromium	...	...	0·4—0·6%
Titanium	...	...	0·25% maximum
Iron ...	...	...	0·5% maximum
Silicon ...	...	...	0·25% maximum
Copper	...	...	0·1% maximum
Manganese	...	...	0·1% maximum
Lead ...	...	...	0·05% maximum
Tin ...	...	...	0·05% maximum
Aluminium	...	...	Remainder

*Registered trade name of the Frontier Bronze Corporation, Niagara Falls, USA.

Influence of Constituent Elements
Zinc and Magnesium

As these two important constituents are increased, so does the tensile strength of age hardened castings increase but at the expense of ductility. Best results appear to be obtained when magnesium is limited to the lower half of the specified range, keeping zinc to the upper half.

Chromium

This element is said to improve stress corrosion resistance but when present to a maximum specified, primary particles may form in the cast structure, having an embrittling effect, so chromium is best kept to the lower range of the specification.

Titanium

Investigations have shown that there is no advantage in exceeding 0·15% titanium. However, adequate grain refinement has been obtained with only 0·03% titanium additions, if boron is also present. Grain refinement with NUCLEANT 2 tablets, plunged into the melt, is therefore recommended, as both titanium and boron are thereby released.

Silicon

This has a harmful effect on tensile properties and should be kept as low as possible.

Required Mechanical Properties

	Sand Cast Test Bar
0·2% proof stress (N/mm²) ...	170
Ultimate tensile stress (N/mm²)	215
Elongation (%)	4

Typical Properties

The outstanding feature of this alloy is its property of age-hardening at room temperature. While artificial ageing at 180°C (356°F) for 10 hours, 48 hours after casting, may be substituted, an initial natural ageing period of 21 days is recommended.

Test Bars

The most suitable form of test bar is the standard E-type bar (as shown on page 152), the moulds for which should be made from an accurate pattern incorporating all the requirements for running and feeding. For diecastings the wedge type metal mould should be used.

Test bars should be cast at 730°C ± 13°C (1346°F ± 24°F).

The following points on the mechanical properties of test bars are given for guidance :

(a) High elongation, low tensile, low yield point—magnesium content to be INCREASED.

(b) Low elongation, high tensile, high yield point—magnesium content to be DECREASED.

(c) Low elongation, low tensile, with or without low yield point, indicates wrong melting or pouring practice. Items to be checked—the test bar pattern and mould, sand moisture, melting and/or pouring temperature, silicon or copper contamination and chromium segregation.

(d) Shiny, star-like fracture of the test bar with low elongation and/or segregation indicates that the metal has not been thoroughly stirred before grain refining, degassing and pouring.

(e) White speckled fracture indicates gassed metal and is due to one, or all, of three things—too slow melting time ; excessive temperature during melting ; smoky combustion of furnace fuels.

(f) Coarse columnar grain extending inward from outside edge of the bar indicates that the metal has been poured too hot, or need for efficient grain refinement of melt.

(g) An oxide film across the fracture indicates that the sand is either too wet or has been rammed too hard or that the metal has been too cold. If the film across the fracture is heavy enough to be almost a misrun, this indicates cold metal. Turbulent metal flow in the mould is another cause of trapped oxide films.

Melting Technique Using a Dry Flux Cover

I. Heat crucible to red heat.

2. Charge in ingot and scrap material, runners, risers, etc.,

bearing in mind the resultant alloy may need a further addition of magnesium.

3. Melt as rapidly as possible and when the charge becomes "pasty" sprinkle COVERAL 11 on the surface, using 500 g per 100 kg. The temperature of the metal should not be allowed to exceed 800°C (1472°F).
4. Grain refining and degassing before casting are essential. Plunge to the bottom of the melt NUCLEANT 2 tablets using 250 g per 100 kg when temperature begins to fall. Keep tablets immersed until bubbling ceases. Then plunge the DEGASER 400 in a similar way, using 300 g per 100 kg. Add magnesium, if necessary, by plunging with a pre-warmed hooked steel rod. Plunging tools, skimmers, etc., should be treated with FIRIT 1 to avoid iron pick-up.
5. At appropriate temperature skim off surface dross and pour.
    NB — Always maintain an unbroken flux cover during melting and if the flux thickens or exposes fresh molten metal apply more COVERAL 11.
    NOTE: The foregoing recommendations are based upon the widely accepted practice of using a dry type of covering flux, of which COVERAL 11 has been specified.

Alternatively, a fluid fluxing technique is advised where the magnesium content of the charge metal is already low. A fluid flux cover such as COVERAL 65 prevents the magnesium content becoming too low. (See Information Sheet 60 for further details.)

Pouring Temperatures

Under 13 mm section	... ...	770°C (1418°F)
13—38 mm section	....	750°C (1382°F)
Over 38 mm section	... ...	730°C (1346°F)

It will be seen that recommended pouring temperatures are about 40°C (104°F) above normal and this is so because the fluidity of Frontier 40E alloy is low at pouring temperatures used for the more common aluminium alloys.

Moulding Techniques

Normal aluminium foundry sands are quite suitable. Moisture in the sand should be around 5—6% maximum, consistent with good moulding practice, as above this figure of moisture there is metal—mould reaction which forms oxide, scum, and other drosses.

To obtain excellent casting finish, green sand moulds should be treated with TERRACOTE 20 or with MOLDCOTE 6 or 9 spirit based types of dressing.

Dry sand moulds and cores should be dressed with a water suspension of TERRACOTE 2 for best quality work or with TERRACOTE 20 for ordinary commercial work.

As Frontier 40E alloy is most susceptible to hot tearing and cracking, moulds should not be so inflexible as to restrain contraction.

Top pouring is very practicable providing it is arranged to minimise turbulence. Isolated heavy sections must be individually risered or chilled but feeding is preferred as the results are more predictable. For feeding extensive heavy sections, the use of excessively large feeders is not advisable because they would take too much room and would retard the cooling of the section with adverse effects. As the shrinkage of Frontier 40E alloy is high compared with the common alloys, ample feeding is necessary particularly during the pasty stage. It is therefore advisable to use FEEDEX or KALMIN sleeves to assist feeding as much as possible.

Diecasting

The fact that Frontier 40E alloy is subject to hot tearing and cracking may have retarded its use in diecasting procedure. Nonetheless quite large and complex diecastings have been produced by using alternative methods of coring such as sand or shell moulded cores. Additionally, a properly designed feeding system can minimise the tendencies to tear and draw.

Owing to the comparatively low fluidity of this particular alloy, it is advisable to cast at a slightly increased temperature, say in the region of 30 Centigrade degrees (54 Fahrenheit degrees) higher than for the more common aluminium alloys.

General Notes

Because of its high strength and shock resistance Frontier 40E alloy is firmly established in the United States of America, notably for the manufacture of services castings and in a wide variety of industrial applications including castings for mining, oil refinery and ventilating equipment, road vehicle parts (e.g., rear axle castings), rail jacks and hand tools.

It is particularly useful as a replacement for heat treated alloys where heat treatment can cause distortion.

MELTING, FLUXING, GRAIN REFINING, ETC.,
AS APPLIED TO MAGNESIUM BASE ALLOYS

General Notes

The primary advantages of magnesium alloys are their lightness combined with strength (they are about 40% lighter than the average aluminium alloy), resistance to fatigue failure and embrittlement in use, and freedom from gas pin-holing and porosity. They are used extensively for aircraft castings, etc. They are also used, but to a more limited extent, in the motor car industry. They are fairly widely used in the heavy vehicle industry.

Aluminium is the main alloying constituent. It increases fluidity and decreases shrinkage, but hot shortness becomes more evident with rising aluminium content. Magnesium alloys, because of their lightness, are easy to set up for machining and have outstandingly good machining properties. However, the cutting edge of the tool must not be allowed to dull, as the heat generated might be sufficient to ignite the chips. The heavier the cut the less chance there is of ignition. Dry machining is recommended, but if cooling is required, a perfectly dry air blast should be used.

They can be sand cast and both gravity and pressure die cast.

Two of the principal casting alloys of this group are L123 and L121. These alloys are also known under the trade names of Elektron AZ91 and Elektron A8. Many other proprietary alloys are also in use.

Chemical Composition

	L123	L121
Aluminium	9·0—10·5%	7·5—9·0%
Zinc	0·3—1·0%	0·3—1·0%
Manganese	0·15—0·4%	0·15—0·4%
Copper	0·15% max.	0·15% max.
Iron	0·05% max.	0·05% max.
Silicon	0·30% max.	0·30% max.
Tin	0·1% max.	0·1% max.
Nickel	0·01% max.	0·01% max.
Magnesium	Remainder	Remainder
Total impurities	0·40% max.	0·40% max.

Typical Mechanical Properties (as cast)

	L123	L121
0·1% proof stress (tons/in²)	4·5	4·5
Ultimate tensile strength (tons/in²)	8·5	10·0
Elongation (% on 2 in.)	1·0	3·0
Brinell hardness number	55	55

Test Bars

Test bars are generally dry-sand moulded to the DTD pattern (see page 136).

Bars should be poured at about 760–780°C. At the start of pouring, the mould should be held at an angle of about 45° from the vertical and then brought gradually to the upright position as the head fills. Pouring should be carried out steadily and over a period of not less than 10 seconds.

Heat Treatment

Although many magnesium alloys containing not less than 6% aluminium respond to heat treatment, the DTD specification calls for L123 and L121 to be supplied in the as-cast condition.

Casting Characteristics

Fluidity	With most alloys excellent—increasing with aluminium content.
Pressure tightness ...	Due to absence of gas porosity pressure tightness is generally very good, but L123 gives particularly good results.
Hot tearing	Hot tearing is not pronounced in most magnesium alloys but may increase with aluminium content.
Patternmakers' shrinkage	1·6%

Melting Considerations

Molten magnesium alloys attack firebrick and refractory furnace linings resulting in harmful silicon contamination and for this reason steel—pressed or cast—crucibles are used. Iron is also slightly soluble in magnesium alloys, but it has a much less harmful effect than silicon. Scrap should be cleaned and if possible shot-blasted to remove any adhering sand as a further precaution against silicon pick-up. To eliminate ladling, the molten alloy should, if possible, be poured direct from the melting pot.

Melting Procedure

1. Heat up the steel crucible and dust the bottom with a little MAGREX 60.

2. Charge in ingot and clean scrap (runners, risers, etc.).

3. Dust the solid charge with a generous quantity of MAGREX 60 using about 500 g per 50 kg of metal.

4. Add remainder of charge and continue melting rapidly, keeping the fluid protective flux cover intact as much as possible.

5. When the melt has reached a temperature of about 750° C stop heating.

6. Skim off the flux remaining on the top surface of the melt and as each patch of metal is exposed dust in more MAGREX 60 until the whole surface is again covered with a fairly thick layer. About 2 kg per 100 kg should be used at this stage.

7. The flux layer will now have melted and should be thoroughly "washed" into the melt, using a plunger (preferably the solid type) in a stirring motion, to bring the fluid flux into contact as much as possible with the molten metal. The plunger should be well coated with a refractory wash of FIRIT and dried thoroughly before use.

8. Stirring should be continued until the metal takes on a bright clean appearance, free from the "frothy" look of partially cleansed metal. As the flux dissolves and entraps oxides it increases in weight and finally sinks to the bottom of the melt,

forming a sludge. If the metal is very dirty a further quantity of MAGREX 60 should be stirred in until the metal takes on a bright lustrous appearance.

9. The cleaned surface should now be dusted with a generous layer of MAGREX 60 so as to curtail burning, which, however, should not be excessive if the temperature is maintained at about 750°C.

10. The sludge may be removed at this stage, but it is recommended that it be left in the bottom of the pot and the metal decanted from it in due course.

11. With the temperature of the metal still around 750°C, grain refinement should now be carried out, using NUCLEANT 200. Using 125 g per 50 kg the required amount is thrown on to the surface of the melt. Plunging is not usually necessary as the tablets are sufficiently heavy to sink to the bottom of the melt, where they decompose and exercise a powerful degassing as well as a grain refining action. In some alloys containing large quantities of other elements such as aluminium, plunging may be necessary, when this degassing action will be most beneficial in reducing harmful micro porosity.

   This method of grain refining is based on the theory of carbon nucleation; NUCLEANT 200 decomposes to form throughout the melt small "specks" of carbon which act as nuclei and around which a very large number of small grains form instead of a lesser number of larger grains, as would occur in a melt perfectly free of such carbon particles or other materials capable of acting as nuclei. It replaces the old method of grain refining by superheating the melt to a temperature of about 850—900°C, and is much more satisfactory. It also shows a marked economy in fuel consumption.

12. During the time the melt is cooling to the correct temperature it should be skimmed and the cleaned area immediately dusted with INERTEX to prevent burning.

13. Pour carefully, dusting the metal stream as it flows into the mould with sulphur. Care must be taken to keep back any slag and particularly when nearing the end of the pour, to prevent any sludge entering the mould.

14. Remove the sludge from the bottom of the pot and thoroughly scrape the sides and bottom before returning it to the furnace for recharging.

Casting Temperatures

Casting temperatures vary with section thickness but generally fall within the following limits :

Light castings—
    under 13 mm section   .   ...   780—810°C

Medium castings—
    13—38 mm section   760°C

Heavy castings—
    over 38 mm section   ...   ...   730°C

Use of Inhibitors

With natural or synthetic sand when used green, it is most important that a chemical inhibitor should be added to the facing sand to prevent the reaction which would occur between the molten metal and the moisture of the sand. Several materials give satisfactory results, but the more generally used are boric acid or ammonium bi-fluoride when used singly, or boric acid with additions of sulphur.

Sulphur must *not* be used in conjunction with ammonium bi-fluoride. The quantity required in each case depends upon the moisture content of the sand and section thickness of the casting, an increase in either calling for an increase in the amount of inhibitor used. Generally, however, 4—6% sulphur with 0·5% boric acid is used or up to 2% of ammonium bi-fluoride alone.

It is recommended that new facing sand should always be used and that all sand used for such magnesium alloys should be segregated from that used for other light alloy work.

Moulding Technique

Magnesium alloys can be satisfactorily green or dry sand moulded in either natural or synthetic moulding sand, the requirements being high permeability and fairly low green and dry strength,

162

but it is more normal to use synthetic sands. Magnesium alloys can be cast easily into moulds produced by the carbon dioxide process and by shell moulding processes.

Sand Properties

Suitable properties of both types of sand are as follows :

		Natural sand	Synthetic sand
Grain fineness (AFA No.)	...	120	80—100
Clay content (%)		10	4
Moisture (%)		5—6	2·5—3·5
Permeability AFA No. (green sand)		30—40	80—100
Green compression strength (lb/in^2)		10—13	11—15

Cores should be made of fairly coarse silica sand bonded with an oil or synthetic resin binder and some agent to give the necessary green bond. The binder should evolve a minimum of gas and absorb a minimum of moisture. Magnesium alloys have appreciable shrinkage and tend to be hot short, thus hard cores must be avoided. It is recommended, therefore, that cores should be well baked and preferably over-baked rather than under-baked.

Mould Dressings

Green sand moulds should be dusted with TERRACOTE 2 or 20 and sleeked to improve casting finish. Alternatively, an excellent finish will be obtained by applying MOLDCOTE 9. This spirit-based dressing is ignited and burns off from the treated surface to leave a firm, closely-adhering coating and gives the effect of skin-drying or torching.

Dry sand moulds and oil sand cores should be dressed with a water suspension of TERRACOTE 2 for best quality work, or with TERRACOTE 20 for ordinary commercial work. Both moulds and cores should, of course, be thoroughly dried before casting.

Running and Gating Methods

As magnesium alloys oxidise rapidly, every effort should be made to ensure non-turbulent pouring and to prevent dross being carried into the mould cavity. Bottom pouring is generally used, often from a ring runner. A shrink bob is frequently incorporated at some point near the gate, particularly if there are thin sections followed by thick ones. The downgate is generally cut rectangular to prevent a swirling effect of the metal, while runners of triangular section showing a larger surface area than the usual circular section runners, help to remove dross. Alternatively, if runners of circular section are used, a skimgate either of baked core sand or perforated metal should be incorporated. The runner should be enlarged slightly where the gate is situated to ensure it will not choke the flow of metal.

As magnesium is so light, a pouring basin should be built up above the top surface of the mould to provide greater hydrostatic pressure. The basin should be roughly oblong in shape, with the mouth of the downgate at one end and a shallow depression at the other, into which the metal is poured. This prevents splashing down the gate into the mould and helps also to trap any dross.

Feeding Methods

Running and feeding methods should be as for a moderately high shrinkage alloy, e.g., fairly large diameter runners and large feeding heads should be used. Running methods should promote directional solidification, i.e., the sections furthest away from the gates and heads should solidify first. Heads should not be "necked"; they should be of the full section they have to feed and should be of such a size as to ensure that they remain liquid until the casting itself has solidified. On heavy sections where casting design prevents the placing of adequate heads, chills should be used. They should be perfectly clean and preferably coated with a dressing such as MOLDCOTE 9 and dried thoroughly before use. Chills made of cast iron and perforated or grooved for venting purposes will be found effective.

Change in section of the castings should be gradual, from heavy to light, and generous fillets of a radius not less than 13 mm should be used.

The necessity for relatively large feeder heads has already been touched on. Quite considerable economies in yield can, however, be obtained from the use of KALMIN 33 or KALMIN 4170 insulating riser sleeves. These resilient, highly permeable, high insulation prefabricated sleeves are inert to the magnesium base alloys, and their use can permit the use of risers smaller in mass than would otherwise be possible. Again heat loss to atmosphere becomes increasingly important as insulation of the riser walls improves and a layer of FEEDOL 17 should be applied to the open surface immediately the riser is full.

Gravity Die Casting

Dies should always be designed with ample venting particularly to "blind" areas. Magnesium alloys are so light in weight that little metallostatic pressure is available to displace mould air and the possibility of short runs or cold shuts is enhanced. Downsprues, for the same reason, are large in cross-sectional area relative to aluminium practice and choke is provided by narrow flat ingates. Feeding heads, as already mentioned, should be of large relative mass. A minimum casting wall thickness will be about 5 mm. Die dressings may be the same as used for aluminium alloys.

The following observations will apply to bale-out furnaces fitted with steel crucibles but where no other special adaptation such as hooding or the provision of a sulphur dioxide atmosphere is afforded. Under these open conditions it is essential that the surface of the metal is protected in order to prevent burning. At the same time the introduction of fresh scrap, ingot, etc., means that oxides can build up and a suitable flux cover should be capable of carrying out both tasks.

Three products are commonly utilised for this purpose, viz:

MAGREX 36 FF.— A fused compound based upon magnesium chloride and supplied in the form of "chippings." At the same time, a heavy ingredient is used as part of the mixture, with the result the slag tends to descend of its own volition through the charge as a whole. The product is therefore what might be termed self-cleansing.

NUCLEANT 200 TABLETS. The term "degasser" is used in the sense of expelling hydrogen, although a major influence of the decomposition products is to nucleate the molten metal with carbon, thus effecting considerable grain refinement without recourse to superheating. Furthermore, as nascent magnesium chloride is also formed within the charge, a natural cleansing influence follows as a matter of course.

INERTEX.—A "dusting" compound which is far more efficient than ordinary flowers of sulphur. The compound is partly volatile and partly "static," the latter feature manifesting itself in the form of a thin but tenacious film which is not easily ruptured. This film is far more readily controlled than that applying to a flux slag, and when baling-out, it is a comparatively simple operation to push back the fused Inertex with the hand ladle, thus allowing the latter to fill with clean magnesium.

Set out below is a recommended melting sequence, using the appropriate products.

1. Heat up the steel crucible.

2. Introduce MAGREX 36 FF "lumps" on a basis of about 3% by weight of the eventual charge capacity.'

3. Add pre-heated magnesium ingot and/or scrap in the form of runners, risers, etc. This should be done in stages until the melt is complete. Care should be taken to retain the flux "cover", any deficiency in this respect being made good by a further introduction of MAGREX 36 FF. Temperature of the charge should be closely controlled with the objective in view of a working range of 730—750 °C.

4. The refining stage has now been reached, the MAGREX 36 FF "wet" slag being thoroughly but gently worked into the bulk of the metal over a period of some 3—4 minutes. This should be carried out until such time as the magnesium assumes a bright lustre. If, after refining, the metal still looks "frothy" then this is an indication it is still in a comparatively dirty condition and requires more refining with a fresh introduction of fluxing compound.

    It will be appreciated that one of the major functions of the fluxing slag is to chemically dissolve and mechanically entrain oxides, plus other non-metallics. This "loaded" slag will therefore tend to cling to the sides of the melting pot, although the bulk of it gradually descends to the bottom. Consequently, atmospheric action upon the relatively bare surface of the melt should be curtailed by adding a little fresh MAGREX 36 FF, thus forming an adequate seal.

5. Grain refinement should now be practised by introducing NUCLEANT 200 tablets at the rate of approximately 0·25% of magnesium base alloy. As the specific gravity of this tabletted compound is of the order of 2·1 they sometimes sink of their own accord dependent upon the density of the particular type of magnesium alloy. However, as plunging is necessary in certain cases, a well-drilled plunging tool should be employed.

This tool, of course, should be coated with a refractory wash and FIRIT has proved very suitable for the purpose. It is important also to make sure the plunger is *thoroughly* pre-heated prior to each application bearing in mind the tool inevitably comes into contact with the "wet" fluxing slag, the latter being very hygroscopic when cold.

The covering slag in the centre of the charge is pushed aside with a skimmer or the plunger itself, and the requisite number of tablets placed on the freshly exposed area. The reaction is by no means violent, decomposition taking place steadily over a period of 1 to 2 minutes.

6. Metal should now be allowed to "settle" for 10 minutes or so (under protective cover of flux).

7. Skim off the surface slag in stages, using a spade type tool for the purpose. As unprotected metal is exposed, the area should be immediately dusted with INERTEX 1, thus preventing any tendency to burn in contact with the atmosphere. We are now left with a thin film of this product and ladling-out can commence. An average pouring temperature is about 750°C.

8. Although the INERTEX film is to a large extent self-healing, it is desirable to keep a quantity of the powder in a drilled canister, thus allowing frequent applications to cover the exposed surfaces.

As suggested earlier, the diecaster merely pushes back part of the surface skin with the bowl of his ladle, so allowing clean magnesium to enter. The hand ladle itself should preferably be of a covered or hooded type, and incorporate a "teapot" spout. This gives a controlled pour into the die cavity and, furthermore, the hood does help to prevent surface burning during the short period involved in transferring metal from bale-out furnace to die.

9. The die cavity itself should, of course, be flushed with sulphur dioxide delivered under pressure from a cylinder prior to each pour.

10. The majority of magnesium gravity diecasting exponents employ their bale-out furnaces on a continuous rather than a batch basis. This means that fresh ingot metal and/or scrap is added from time to time to maintain reasonable metal capacity. Such additions, of course, should not be made too frequently, as the general disturbance could lead to the presence of flux inclusions. Consequently, wherever conditions permit, topping-up should be done at prescribed intervals, refining and grain refining following in sequence as already laid down.

11. With regard to the spent flux in the form of "sludge" which accumulates around the sides and at the bottom of the steel crucible, it is customary to remove this on, say, two occasions in an 8 hour shift. This involves thoroughly scraping the walls of the pot with a skimming tool, followed by a suitably shaped tool for removing sludge from the bottom. A lot depends upon the amount of metal which has melted in a certain period, some firms run through a complete shift leaving sludge removal as the last job of the day.

To prevent sticking, to give improved surface finish and — where necessary — to give control over the rate of cooling, the use of DYCOTE is essential. DYCOTE 140 is particularly suitable for use with magnesium alloys. It has especially good insulating properties to prevent too severe a chilling effect of the die and gives a first-class casting finish.

Pressure Die Casting

Because of the choking of the nozzle due to the disturbance of the flux cover with consequent dross formation, the conventional "gooseneck" type of machine cannot be used. The cold chamber machine, however, gives completely satisfactory results. With the cold chamber machine a pressure of about 4000 lb/in^2 forces the metal at a little above its freezing temperature into the die cavity. Both these factors help the production of strong, sound castings free from porosity and with an excellent finish.

Pressure die castings can be made with a wall thickness of less than 3 mm. A small addition of zinc improves fluidity and surface finish but as it tends also to increase hot shortness it should not exceed about 0·6%.

Melting Magnesium Scrap, Swarf, Turnings, etc.

A low fusion-point flux is necessary in order to protect the large surface area of the charge against oxidation and also to absorb dirt and non-metallics. The appropriate flux in this case will be MAGREX 60.

The charge should be commenced by pre-fusing MAGREX 60 to the extent of 5—10% by weight of the charge — the precise quantity must depend on the condition of the scrap and swarf.

The pre-heated scrap or dried swarf should be introduced into this flux bath in relatively small quantities.

The aim at all times should be to preserve a complete layer of molten flux, otherwise the losses by oxidation will be extremely high.

When the charge is complete and at a temperature of approximately 700—750°C add fresh MAGREX 60 to the extent of 2% by weight of the molten metal. The fused flux should then be thoroughly stirred into the melt for a period of from 3 to 5 mins, using a coated iron bar for this purpose — the bar is suitably bent to enable the operator to "whisk" the metal and flux, thus bringing the two into as intimate contact as possible. During the operation, most of the flux will disappear and form a "sludge" on the bottom of the crucible.

This sludge may be removed immediately by dredging with a suitable tool or the metal may be decanted.

The first method should be carried out as follows:

The sides of the crucible should be well scraped and the sludge finally dredged out, using special "spoon" tools for the purpose. The metal should then be ladled into the ingot moulds, and the necessary protection for the metal remaining in the pot can be maintained with INERTEX. A dusting of INERTEX on the solidifying magnesium ingots will curtail undue burning.

If it is found more convenient to remove the residual flux after pouring, it is necessary to carefully watch that a few inches of magnesium are left in the crucible. This residue of metal and sludge should be poured into separate moulds followed by final and thorough cleaning of the crucible.

MELTING AND RECOVERY OF ALUMINIUM SWARF AND SMALL SCRAP SUCH AS SHEET CLIPPINGS, MILK BOTTLE TOPS, etc.

Rotary type furnaces are favoured. The flame from the burner strikes the furnace lining and the exhaust spreads around the furnace walls which absorb heat. Then as the furnace rotates this heat is taken up by the metal. Melting must take place under a cover of molten flux, otherwise heavy oxidation results, as each piece of swarf, as it melts, forms a globule of liquid surrounded by a shell of oxide. The skin tension of this oxide prevents coalescence and in consequence of the large surface area presented by the mass of globules, increase of oxide formation and loss of yield is bound to take place. Therefore, the skin of oxide has to be ruptured in order to allow coalescence. The molten flux encourages this by chemical action, while the rotation of the furnace provides a mechanical action. COVERAL 48 or 57 are appropriate fluxes melting at 605 and 590°C respectively.

Procedure for 500 kg Furnace

1. Charge 100 kg of flux before introducing the turnings, and heat to 850—900°C.

2. This takes 15—20 minutes, after which the furnace is turned over to the charging position and 30 kg of swarf or smaller scrap fed in.

3. Rotation of the furnace turns the swarf in under the flux. The heat from the flux is rapidly transferred to the metal until a state of balance is reached.

4. Consequent cooling of the flux causes it to become pasty, thus affording a complete cover for the melt.

5. After adding the 30 kg of swarf, the furnace is returned to the firing position and when the flame, caused by oil burning from the swarf, dies the burner is re-started and in from 4 to 5 minutes the temperature is back to 750°C. This temperature should not be greatly exceeded in order to keep gas absorption to a minimum.

6. Repeat the process until the balance of the 500 kg of swarf is fed into the furnace. During the latter stages an increased rate of charge of about 45 kg a time can be used. The complete melt takes about 2¼ hours from start to finish.

After the whole 500 kg has melted the flux becomes more viscous through taking up oxides and hydrocarbons from the burnt oil and other impurities. This enables the molten metal to be poured from underneath the flux.

In order to bring back the fluidity of the flux, about 25 kg is removed from the furnace and an equal quantity of fresh flux added. In this way five charges can be completed before a complete change of flux is needed. This brings the flux/metal ratio to 10%. Average reclamation yield is approximately 90%. The recovered metal is degassed and drossed off with COVERAL II. If magnesium has to be removed, the melt is treated with LOMAG 4 (or LOMAG 6 for crucibles). 0.5 kg of LOMAG removes approximately 0·1% of magnesium from a 50 kg melt.

GRAIN REFINEMENT OF ALUMINIUM ALLOYS WITH TITANIUM AND BORON

Both titanium and boron will grain refine or nucleate all the aluminium alloys. They may be added either as the element or as hardener alloys or by direct reduction of titanium and boron containing salts. Hardener alloys have some disadvantages in that considerable amounts of oxide are added simultaneously and slow solution rates often mean areas of high concentration leading to grain boundary borides, etc. The salts method gives good dispersion, but efficiency of pick-up may sometimes be disappointing if conditions are not exactly right.

Considerable experimental work on the individual elements and on combinations has produced results that are often indecisive and inconclusive. On the one hand, it is considered that with the exception of boron, which is assumed to form aluminium boride, the other well-known nucleating agents, e.g. titanium columbium, cerium, zirconium, etc., all react with any available carbon to form the appropriate carbide. This is said to be the important refining agent as distinct from the metal itself. On the other hand, the view is also held that grain refinement consequent upon the addition of titanium and boron is due to the formation of titanium boride as distinct from duplex nucleation from titanium carbide and aluminium boride.

It is now known that single introduction of titanium, added either as hardener or salt to an aluminium alloy, grain refines due to the formation of titanium carbide. An addition of boron results in the formation of nuclei of aluminium boride which grain refines in a similar way. A combined introduction of titanium and boron results additionally in the formation of nuclei of titanium boride and it is generally conceded that boron and titanium added together have a more marked refining effect than when either is added alone.

From all this background it is clear that a combined addition of titanium and boron is desirable and other considerations indicate that a reduction from the appropriate salt would be the best method. These factors led to the development of NUCLEANT 2 tablets the use of which is now standard practice in many countries where consistent quality in the aluminium produced is of prime importance. In appropriate quantities the product is used for sand

and gravity diecastings, forging and extrusion billets, slabs, etc., produced by all conventional processes including semi- and fully continuous methods.

Microspecimens invariably show small and uniform grain structures with columnar tendencies, so often associated with large equi-axial "banding", completely suppressed. At the same time local concentrations of boron, giving rise to grain boundary constituents are avoided. NUCLEANT 2 tablets utilise a titanium: boron ratio of 6:1, which is generally accepted as being the most effective and the standard recommendation is 0·25% of product by weight with metal temperature approximately 750°C. With large capacity melting and holding furnaces where application efficiency can be improved, quite considerable economies in product application rate can be effected.

While NUCLEANT 2 has been developed to meet the major requirement, for other purposes it is possible that nucleating agents based on either titanium or boron alone may be required. These can be met with NUCLEANT 6 (titanium only) or NUCLEANT 186A (boron only).

For the very large user where individual furnaces often contain many tonnes of aluminium alloy there is a further development known as TILITE 101. This is a lustrous grey coloured block weighing 0·5 kg and containing 40% titanium and 4% boron. The highly compressed block is self-sinking in molten aluminium and application is therefore simply a matter of throwing the requisite number of blocks into the cleaned melt. The blocks will dissolve rapidly a matter of 15—30 seconds depending on metal temperature — and a gentle stir is all that is required to ensure good dispersion of the nuclei.

Depending therefore on melt size and on type of nucleating agent it is possible to select a suitable grain refiner to give the degree of refinement required in the macrostructure.

SECTION III

NON-FERROUS CASTING ALLOYS

THE ORDER OF ALLOYING

The table below shows the order in which various elements are added in the alloying of copper and nickel alloys.

The base metal with which the melt commences is usually marked as "1," whilst the various elements added afterwards are marked under their respective heading with the number giving their stage of addition.

Example: GUNMETAL. 88% copper, 10% tin, 2% zinc.

Charge crucible with "1" copper, then add "2" tin, afterwards "3" zinc.

This table applies to virgin metals. If scrap is used in addition this is usually added together with "1."

	Copper	Tin	Zinc	Lead	Nickel	Phosphorus	Remarks
Brass	1	—	2	—	—	—	Stir very well after zinc has been added
Naval brass ...	1	*2	3	—	—	—	Stir very well after zinc has been added
Gunmetal ... copper, tin, zinc, such as 88/10/2	1	*2	3	—	—	—	Stir well before pouring
Gunmetal ... with lead such as 85/5/5/5.	1	*2	4	3	—	—	Stir very well before pouring
Bronzes ... such as 90% copper, 10% tin	1	*2	—	—	—	—	Stir well before pouring
Phosphor bronze	1	3	—	—	—	**2 **4	Stir well
Nickel silver or white metal	***2	3 if any	5	5	***1	—	Stir very well
Cupro nickel or nickel bronzes	***2	—	—	—	***1	—	Stir well

*Deoxidise well with deoxidising tubes DS before "2" is added.
**Phosphorus is best added as phosphor copper. Half the total amount before "3" is added, the other half afterwards.
***Charge the total amount of nickel, then add as much of the copper as crucible will hold. Then commence melting and continue to add copper as melt proceeds in the usual fashion.

BRITISH STANDARD SPECIFICATIONS FOR THE PRINCIPAL NON-FERROUS CASTING ALLOYS

BS 1400·1969 is the first revision of the specification in metric terms. At the same time, the alloys covered have been grouped into three categories according to the approximate degree of usage and availability; There are some additions, some revisions, and some deletions with respect to previous editions of the BS 1400 Specification.

The main alloy groups are as follows:

Group A.　Alloys in common use. PB4, LPB1, LB2, LB4, LG2, LG4, SCB1, SCB3, SCB6, DCB1, DCB3, PCB1.

Group B.　Special purpose alloys. HCC1, CC1—WP, PB1, PB2, CT1, LB5, LG1, AB1, AB2, CMA1, CMA2, HTB1, HTB3.

Group C.　Alloys in limited production. LB1, G1, G3, G3-WP, SCB4.

Tables giving chemical composition, mechanical properties, density, and coefficient of thermal expansion of these alloys are given on the following pages.

Elements	PB4 min %	PB4 max %	LPB1 min %	LPB1 max %	LB2 min %	LB2 max %	LB4 min %	LB4 max %	LG2 min %	LG2 max %	LG4 min %	LG4 max %
Designation:	PB4		LPB1		LB2		LB4		LG2		LG4	
Material:	Phosphor bronze (copper—tin—phosphorus)		Leaded phosphor bronze		80/10/0/10 leaded bronze		85/5/0/10 leaded bronze		85/5/5/5 leaded gunmetal		87/7/3/3 leaded gunmetal	
Copper	REM		REM		REM		REM		REM		REM	
Tin	9·5	—	6·5	8·5	9·0	11·0	4·0	6·0	4·0	6·0	6·0	8·05
Zinc	—	0·5	—	2·0	—	1·0	—	2·0	4·0	6·0	1·5	3·0
Lead	—	0·75	2·0	5·0	8·5	11·0	8·0	10·0	4·0	6·0	2·5	3·5
Phosphorus	0·4	—	0·3	—	—	0·10	—	0·10	—	—	—	—
Nickel	—	0·5	—	1·0	—	2·0	—	2·0	—	2·0	—	2·0§
Iron	—	—	—	—	—	0·15*	—	—	—	—	—	0·20*
Aluminium	—	—	—	—	—	0·01*	—	—	—	0·01*	—	0·01*
Manganese	—	—	—	—	—	—	—	—	—	—	—	—
Antimony	—	—	—	—	—	0·5	—	0·5	—	—	—	0·25*
Arsenic	—	—	—	—	—	—	—	—	—	—	—	—
Iron + arsenic + antimony	—	—	—	—	—	—	—	—	—	0·50*	—	0·40*
Silicon	—	—	—	—	—	0·02*	—	0·02*	—	0·02*	—	0·01*
Bismuth	—	—	—	—	—	—	—	—	—	0·05*	—	0·05*
Total impurities	—	0·50	—	0·50	—	0·50	—	0·50	—	0·80	—	0·70

Total impurities include elements marked *

†DCB1 — 0·1% lead if required.
‡DCB3 — nickel to be counted as copper.
§ Tin + ½ nickel content 7·0—8·0%.

178

Elements	SCB1 min. (%)	SCB1 max. (%)	SCB3 min. (%)	SCB3 max. (%)	SCB6 min. (%)	SCB6 max. (%)	DCB1† min. (%)	DCB1† max. (%)	DCB3‡ min. (%)	DCB3‡ max. (%)	PCB1 min. (%)	PCB1 max. (%)
Designation: Material:	Brass for sand casting		Brass for sand casting		Brass for brazable casting		Brass for die casting		Brass for die casting		Brass for pressure die casting	
Copper	70·0	80·0	63·0	70·0	83·0	88·0	59·0	63·0	58·0	63·0	57·0	60·0
Tin	1·0	3·0	—	1·5					—	1·0	—	0·5
Zinc	REM		REM		REM		REM		REM		REM	
Lead	2·0	5·0	1·0	3·0	—	0·5	—	0·25	0·5	2·5	0·5	2·5
Phosphorus	—	—	—	—	—	—	—	—	—	—	—	—
Nickel	—	1·0	—	1·0	—	—	—	—	—	1·0	—	—
Iron	—	0·75	—	0·75	—	—	—	—	—	0·8	—	0·3
Aluminium	—	0·01´	—	0·1*	—	—	—	0·5	0·2	0·8	—	0·5
Manganese	—	—	—	—	—	—	—	—	—	0·5	—	—
Antimony	—	—	—	—	—	—	—	—	—	—	—	—
Arsenic	—	—	—	—	0·05	0·20	—	—	—	—	—	—
Iron + arsenic + antimony	—	—	—	—	—	—	—	—	—	—	—	—
Silicon	—	—	—	—	—	—	—	—	—	0·05´	—	—
Bismuth	—	—	—	—	—	—	—	—	—	—	—	—
Total impurities	—	1·0	—	1·0	—	1·0 (incl. Pb)	—	0·75	—	2·0 (excl. Ni + Pb + Al)	—	0·5

179

CHEMICAL COMPOSITION

Designation	HCC1†		CC1-WP‡		PB1		PB2		CT1		LB5	
Material	High conductivity copper		Copper chromium		Phosphor bronze (copper–tin–phosphorus)		Phosphor bronze (copper–tin–phosphorus)		Copper–tin		75/5/0/20 leaded bronze	
Elements	min. (%)	max. (%)	min. (%)	max. (%)	min (%)	max. (%)	min. (%)	max. (%)	min. (%)	max. (%)	min. (%)	max. (%)
Copper			REM		REM		REM		REM		REM	
Tin	Castings		—	—	10.0	—	11.0	13.0	9.0	11.0	4.0	6.0
Zinc	shall be		—	—	—	0.05*	—	0.30	—	0.05*	—	1.0
Lead	made from		—	—	—	0.25*	—	0.50	—	0.25*	18.0	23.0
Phosphorus	copper		—	—	0.50	—	0.15	—	—	0.15	—	0.10**
Nickel	complying		—	—	—	0.10*	—	0.50	—	0.25*	—	2.0
Iron	with		—	—	—	0.10*	—	0.15*	—	—	—	—
Aluminium	BS 1035–		—	—	—	0.01*	—	0.01*	—	—	—	—
Manganese	1037.		—	—	—	—	—	—	—	—	—	—
Antimony	Added		—	—	—	—	—	—	—	—	—	0.50
Arsenic	elements		—	—	—	—	—	—	—	—	—	—
Iron + arsenic + antimony	and impurities not specified		—	—	—	—	—	—	—	—	—	—
Silicon			—	—	—	0.02*	—	0.02*	—	—	—	0.01**
Bismuth			—	—	—	—	—	—	—	—	—	—
Magnesium			—	—	—	—	—	—	—	—	—	—
Chromium			0.60	1.2	—	—	—	—	—	—	—	—
Total impurities			—	—	—	0.60	—	0.20	—	0.80	—	0.30

†HCC1 —maximum resistivity = 0·019 microhm-metre (approximately 90% IACS).
‡CC1-WP—maximum resistivity = 0·022 microhm-metre (approximately 80% IACS).

NOTE. Copper which has a resistivity of 0·017 241 microhm-metre is said to have a conductivity of 100% IACS

§AB1 and AB2—when castings are to be welded lead not to exceed 0·01%.
‖HTB1—subject also to microstructure requirements.
¶Material was designated 30 ton and 48 ton respectively.

OF GROUP B CASTINGS

Designation	LG1		AB1§		AB2§		CMA1		CMA2		HTB‖		HTB3	
Material	83/3/9/5 leaded gunmetal		Aluminium bronze (copper–aluminium)		Aluminium bronze (copper–aluminium)		Copper–manganese aluminium		Copper–manganese aluminium		46 hbar¶ high tensile brass		74 hbar¶ high tensile beta brass	
Elements	min. (%)	max. (%)	min. (%)	max. (%)	min. (%)	max. (%)	min. (%)	max. (%)	min. (%)	max. (%)	min. (%)	max. (%)	min. (%)	max. (%)
Copper	REM		REM		REM		REM		REM		55·0	–	55·0	–
Tin	2·0	3·5	–	0·1**	–	0·1**	–	1·0	–	1·0	–	1·0	–	0·20
Zinc	7·0	9·5	–	0·50	–	0·50	–	0·50	–	0·50	REM		REM	
Lead	4·0	6·0	–	0·05**	–	0·05*	–	0·05*	–	0·05*	–	0·50	–	0·20
Phosphorus							–	0·05*	–	0·05*				
Nickel	–	2·0	–	1·0	–	1·0	1·5	4·5	1·5	4·5	–	1·0		
Iron			1·5	3·5	4·0	5·5	2·0	4·0	2·0	4·0	0·7	2·0	1·5	3·25
Aluminium	–	0·01**	8·5	10·5	8·8	10·0	7·5	8·5	8·5	9·0	0·5	2·5	3·0	6·0
Manganese			–	1·0	–	1·5	11·0	15·0	11·0	15·0	–	3·0	–	4·0
Antimony	–		–		–		–		–		–		–	
Arsenic	–		–		–		–		–		–		–	
Iron + arsenic + antimony			–	0·75**										
Silicon	–	0·02**	–	0·25**	–	0·1*	–	0·15*	–	0·15*	–	0·10*	–	0·10*
Bismuth	–	0·10**												
Magnesium			–	0·05**	–	0·05*								
Chromium	–		–		–		–		–		–		–	
Total impurities	–	1·0	–	0·30	–	0·30	–	0·30	–	0·30	–	0·2	–	0·2

Analysis must include all elements marked * but not necessarily those marked **.
Total impurities include elements marked * and **

CHEMICAL COMPOSITION OF GROUP C CASTINGS

Designation	LB1		G1		G3		G3WP		SCB4	
Material	76/9/0/15 leaded bronze		88/10/2 gunmetal		Nickel gunmetal		Nickel gunmetal fully heat treated		Naval brass for sand casting	
Elements	min. (%)	max. (%)	min. (%)	max. (%)	min. (%)	max. (%)	min. (%)	max. (%)	min. (%)	max. (%)
Copper	REM		REM		REM		REM		60.0	63.0
Tin	8.0	10.0	9.5	10.5	6.5	7.5	6.5	7.5	1.0	1.5
Zinc	—	1.0	1.75	2.75	1.5	3.0	1.5	3.0	REM	
Lead	13.0	17.0	—	1.5	0.10	0.50	0.10	0.50	—	0.5
Phosphorus	—	0.1**	—	—	—	0.02	—	0.50	—	—
Nickel	—	2.0	—	1.0	5.25	5.75	5.25	5.75	—	—
Iron	—	—	—	0.15*	—	—	—	—	—	—
Aluminium	—	—	—	0.01*	—	0.01*	—	0.01*	—	0.01**
Manganese	—	—	—	—	—	0.20*	—	0.20*	—	—
Antimony	—	0.5	—	—	—	—	—	—	—	—
Arsenic	—	—	—	—	—	—	—	—	—	—
Iron + arsenic + antimony	—	—	—	0.20*	—	0.20*	—	0.20*	—	—
Silicon	—	0.02**	—	0.02*	—	0.01*	—	0.01*	—	—
Bismuth	—	—	—	0.03*	—	0.02*	—	0.02*	—	—
Total impurities		0.30		0.50		0.50		0.50		0.75

182

N.B. Analysis must include all elements marked * but not necessarily those marked **. Total

COLOUR CODE FOR INGOTS

Designation	Colour code	Designation	Colour code
Group A		**Group B**	
PB4	Black/red	PB1	Yellow
LPB1	Black	PB2	Yellow/red
LB2	White	CT1	Black/aluminium
LB4	White/green	LB5	White/brown
LG2	Blue	LG1	Blue/red
LG4	Blue/brown	AB1	Aluminium
SCB1	Green/blue	AB2	Aluminium/green
SCB3	Green	CMA1	Aluminium/red
SCB6	Green/brown	CMA2	Aluminium/yellow
DCB1	Yellow/blue	HTB1	Brown
DCB3	Yellow/brown	HTB3	Brown/red
PCB1	White/blue		
		Group C	
		LB1	White/black
		G3	Blue/black
		SCB4	Green/yellow
		G1	Red

MECHANICAL PROPERTIES OF TEST SAMPLES FOR CASTINGS

Desig-nation	Tensile strength				0·2% proof stress				Elongation on $5·65\sqrt{S_0}$			
	Sand min. hbar‖	Chill min. hbar‖	Con-tin-uous min. hbar‖	Centri-fugal§ min. hbar‖	Sand min. hbar‖	Chill min. hbar‖	Con-tin-uous min. hbar‖	Centri-fugal§ min. hbar‖	Sand min. %	Chill min. %	Con-tin-uous min. %	Centri-fugal§ min. %
Group A alloys												
PB4	19	27	33	28	10*	14*	16*	14*	3	2	7	4
LPB1	19	22	27	23	8*	13*	13*	13*	3	2	5	4
LB2	19	22	28	23	8*	14*	16*	14*	5	3	6	5
LB4	16	20	23	22	6*	8*	13*	8*	7	5	9	6
LG2	20	20	27	22	10*	11*	10*	11*	13	6	13	8
LG4	25	25	30	25	13*	13*	13*	13*	16	5	13	6
Group B alloys												
CC1WP†	—	—	—	—	—	—	—	—	—	—	—	—
PB1	22	31	36	33	13*	17*	17*	17*	3	2	6	4
PB2	22	27	31	28	13*	17*	17*	17*	5	3	5	3
CT1	23	27	31	28	13*	14*	16*	14*	6	5	9	6
LB5	16	17	19	19	6*	8*	10*	8*	5	5	8	7
LG1	18	18	—	—	8*	—	—	—	11	2	—	—

AB1	50	54	—	56	17*	20*	—	20*	18	18	—	20
AB2	64	65	—	67	25	25	—	25	13	13	—	13
CMA1	65	67	—	—	28	31	—	—	18	27	—	—
CMA2	74	—	—	—	38	—	—	—	9	—	—	—
HTB1	47	50	—	50	17	21	—	21	18	18	—	20
HTB3	74	—	—	74	40	—	—	40	11	—	—	13
Group C alloys												
LB1	17	20	23	22	8*	13*	13*	13*	4	3	9	4
G1	27	23	30	25	13*	13*	14*	13*	13	3	9	5
G3	28	—	34	—	14*	—	17*	—	16	—	18	—
G3WP‡	43	—	43	—	28*	—	28*	—	3	—	3	—

*For information only.
†The minimum hardness of CC1WP after heat treatment shall be 100 HB
‡The minimum hardness of G3WP after heat treatment shall be 160 HB
§Values apply to samples cut from centrifugal castings made in metallic moulds. Minimum properties of centrifugal castings made in sand moulds are the same as for other sand castings.
∥1 hbar = 10 MN/m² = 10 N/mm²

DENSITY AND COEFFICIENT OF THERMAL EXPANSION

Alloy designation	Density (g/cm³)	Linear coefficient of thermal expansion x 10⁻⁶ 0–250°C (per °C)
Group A		
PB4	8·8	18·3
LPB1	8·8	18·0
LB2	9·0	18·6
LB4	9·0	18·4
LG2	8·8	18·0
LG4	8·8	18·2
SCB1	8·5	19·3
SCB3	8·4	20·2
SCB6	8·6	18·6
DCB1	8·3	20·8
DCB3	8·3	20·8
PCB1	8·3	20·8
Group B		
HCC1	8·9	17·3
CC1	8·85	17·3
PB1	8·8	18·3
PB2	8·8	18·5
CT1	8·8	18·1
LB5	9·2	18·9
LG1	8·8	17·8
AB1	7·6	17·0
AB2	7·6	17·0
CMA1	7·5	18·6
CMA2	7·5	18·6
HTB1	8·3	20·5
HTB3	7·9	20·5
Group C		
LB1	9·1	18·7
G1	8·8	18·3
G3	8·8	18·3
SCB4	8·3	20·8

THE IMPORTANCE OF FURNACE ATMOSPHERE
WHEN MELTING COPPER BASE ALLOYS

In metal melting one of three types of atmosphere may be experienced, depending on the condition of melting. By atmosphere is meant the space immediately above the metal being melted and this, of course, means inside the crucible or furnace being used.

This atmosphere can be either: (a) neutral, (b) oxidising, or (c) reducing, and upon this and the nature of the metal, will depend whether the metal is good, bad, or indifferent.

If the Atmosphere is Neutral

This means that the atmosphere is not reacting either for or against the metal and that the coke or fuel must be burning completely without excess of oxygen. A neutral atmosphere is therefore obviously a rarity.

If the Atmosphere is Reducing

This means that the fuel is burning with insufficient oxygen for complete combustion and is therefore generating gases which will react with additional oxygen and burn to complete combustion. These are known as reducing gases. If they are soluble in the molten metal, trouble can be expected through gassy metal. These gases are chiefly carbon monoxide, hydrogen, and gaseous compounds of carbon with hydrogen.

If the Atmosphere is Oxidising

This means that the fuel is burning with an excess of oxygen and the conditions are said to be oxidising. In the case of complete combustion, the gases, water vapour and carbon dioxide are formed which, with the addition of some excess oxygen, will be free to affect the metal. In this case oxidation will take place and must be removed by subsequent deoxidation.

Gases which may occur during the melting process from the combustion of the fuel are :—

Carbon monoxide	Hydrogen	Water vapour
Carbon dioxide	Hydrocarbon	

Those which are extremely detrimental to the melt are hydrogen, carbon monoxide, and the hydrocarbons. These are all reducing gases and therefore will create a reducing atmosphere. In the case of brasses, bronzes, copper and nickel alloys, etc., these are best avoided by using an oxidising flux, thus creating an oxidising atmosphere around the metal; the excess oxygen is then removed by subsequent deoxidation.

It will be seen that control of furnace atmosphere, by adjustment of fuel:air ratio and by the use of the correct flux in contact with the metal, will give direct control of metal quality in respect of ingress of harmful gases. The importance of correct furnace operation and careful choice of flux cover is clear if reliable, consistent results are required.

MELTING, FLUXING, AND DEGASSING PROCEDURE

Brasses, Bronzes, and Gunmetals

Copper-based alloys which contain relatively high percentages of zinc, such as the brasses, are normally free from porosity trouble caused by dissolved hydrogen. This is because the vapour pressure of the zinc at the melting temperatures involved is effective in restricting hydrogen absorption. Zinc also acts as a deoxidising agent; therefore molten brass does not need the addition of deoxidisers for the removal of oxygen or oxides. However, the introduction of a small proportion of phosphorus is valuable as a means of increasing fluidity.

For straight copper-zinc alloys the CUPRIT range of fluxes is recommended. Where aluminium forms a part of the alloy, as in manganese bronze and aluminium bronze, it is advisable to employ a flux capable of dissolving and removing oxide of aluminium such as ALBRAL. The silicon bronzes also benefit from treatment with ALBRAL flux.

An entirely different set of circumstances are experienced in the melting of bronzes and gunmetals. These alloys contain a high ratio of copper and little or no zinc so that gas porosity due to hydrogen pick-up becomes more of a possibility.

Hydrogen can enter the melt in a variety of ways such as from atmosphere, corrosion products on scrap, wet furnace tools, damp ladle linings, moisture in fluxes, and so on. The complete combustion of 5 litres of fuel oil results in the formation of about 6 kg of water vapour which may come into direct contact with the surface of the melt.

If water vapour remained as such, little difficulty would follow, for it is doubtful if water vapour is absorbed. Unfortunately, it tends to split up into hydrogen and oxygen at molten metal temperatures and the gases so formed are atomic or nascent and very active. Although the reaction is reversible, in practice equilibrium is set up wherein the amount of dissociation balances recombination. At any given temperature a mixture of hydrogen, oxygen and water vapour is present, the proportions depending on the temperature. As the temperature rises more hydrogen and oxygen are formed. The nascent oxygen immediately unites with metallic constituents to form oxides, whilst the hydrogen dissolves in the melt. So both are removed from the atmosphere above the alloy, which results in the dissociation of more water vapour, giving rise to more atomic hydrogen and oxygen, which in turn will be dissolved. Thus the build-up of gas in the metal continues.

In order to protect molten bronze from attack in this way, it is necessary to provide an oxidising atmosphere during melting. This is achieved by the use of CUPREX fluxes in either block or powder form. The flux is placed in the required ratio of 1% of the charge weight into the base of the preheated crucible, followed by the charge. Then whilst melting proceeds oxidising gases are evolved which bubble up through the melt, thereby acting as an effective barrier against hydrogen pick-up. Towards the latter stages of melting the block fuses and rises to the surface to form a protective layer.

Finally, just prior to pouring, the flux layer is skimmed off and any surplus oxygen removed by plunging deoxidising tubes into the melt. This method of treatment is known as the oxidation-deoxidation technique and is widely recognised as the best procedure for bronzes and gunmetals.

Degassing Copper Based Alloys

For important castings and those which must meet stringent inspection by X-rays and other non-destructive methods, a separate degassing treatment may be advisable. This can be accomplished by plunging and thoroughly scrubbing specialised fluxes into the melt. These REGENEX fluxes are classed as "oxidising" for the gunmetal, bronze, and nickel alloy group, and "neutral" for phosphor bronze and other alloys containing easily oxidisable constituents.

Alternatively, LOGAS 50 briquettes may be plunged into the melt with a suitably protected plunging tool. These briquettes evolve inert gases at the temperatures involved, and degassing is accomplished by a process of diffusion of hydrogen from the metal into the rising bubbles. The process may be used with any copper-based alloy including high conductivity copper, does not affect metal composition, and evolves no fume. However, some cooling of the metal is unavoidable, and crucibles cannot be completely filled with metal because of the necessity to accommodate plunger displacement.

High Lead Bronzes

High lead bronzes, for the purpose of these notes, are considered to be those containing 10% or more of lead. Alloys of copper containing up to 35% of lead form a homogeneous liquid melt above 1083°C. During cooling, solidification begins and copper separates out at temperatures which vary between 1080° and 954°C as the lead varies from 0 up to 36%. On further cooling from 954° to 326°C, the copper and lead exist as separate constituents. The size and shape of the copper and lead particles are of great practical importance and are affected, amongst other things, by the rate of cooling. When solidification is very slow, the lead has time to coalesce into larger "lakes," while the matrix is still in the pasty condition. If, however, solidification is very rapid the lead is entrapped in a fine network of primary copper, which therefore controls the size of the particles.

The high lead bronzes with lead above 36% behave differently from those with a lead content below that figure. The copper and lead, though both molten at 954°C, do not form a homogeneous liquid, but exist largely in the form of fine droplets as an emulsion. As the liquid particles of lead are heavier, they tend to sink to the bottom of the melt. The lower the temperature (above the melting point), the greater is the tendency, which is very marked below about 1000°C. However, the melt can be brought into a uniform and fairly stable emulsion by vigorous agitation. On the other hand, the two metals will form a homogeneous liquid if the temperature exceeds 1325°C. If the rate of cooling from this temperature is sufficiently rapid, then the lead is retained in a fine state of division.

High lead bronze should be melted down under PLUMBRAL flux. Use ½% of the charge weight of PLUMBRAL added with the charge materials or when they commence to melt. Superheat rapidly to the required temperature and then some 5 minutes before pouring, slowly plunge another ½% by weight of PLUMBRAL. This operation is best carried out with a graphite tool or if an iron or steel plunger is used it should be coated previously with a dressing of DYCOTE 36, and thoroughly dried before use.

Small percentages of nickel are an aid to preventing segregation of the lead, especially when more than 20% of lead is present. An addition of 2% DENSITE 50 to the original charge in the crucible prior to melting will be instrumental in providing about 1% of nickel. When nickel is added the pouring temperature should be raised by about 20°C. With very high lead contents the introduction of sulphur also exercises an important influence on keeping the lead dispersed in a finely divided form.

Suggested pouring temperatures are·

15% lead	...	...	1080–1100°C
20% lead	...	...	1050–1070°C
30% lead	...	...	1030–1050°C

MELTING AND FLUXING PROCEDURE FOR HIGH CONDUCTIVITY AND COMMERCIAL COPPER CASTINGS

High Conductivity Copper

The production of castings from pure copper presents many difficulties, consequently, even pure copper castings seldom have conductivities higher than 80% as measured electrically in International Annealed Copper Standard units. Copper in the annealed condition, which has a specific electrical resistance of $1\cdot72$ microhms per cubic centimetre at $20°C$, is said to have 100% electrical conductivity IACS.

It is often more convenient to refer to the relative electrical conductivity of copper rather than to its resistance, and such conductivity is expressed as a percentage of that of annealed copper based on the international standard. This standard does not represent the highest possible conductivity and it is possible to obtain copper having a conductivity of 1% or even up to 2% higher than that taken as the standard.

Thermal conductivity is also an important property which has to receive major consideration for some applications. It is measured in calories per second per square centimetre per degree centigrade. Generally speaking, purely thermal applications are less critical and lower copper contents (99·50% min.) are common. Thus, any copper casting produced with a view to the development of maximum electrical conductivity will automatically satisfy the requirements of thermal conductivity as well.

Many of the impurities likely to be present in copper lower its electrical conductivity seriously. The commonly found impurities are residues from the deoxidation treatment which are ordinarily soluble in copper and mostly have the effect of depressing the conductivity values. Conductivity is decreased by any element which enters into solid solution with copper. Even silver which has a higher conductivity than copper will reduce the overall conductivity slightly when added to it. The accompanying list gives some idea of the influence of a number of elements, which are commonly present as impurities, in lowering the electrical conductivity.

0·1%	Aluminium	lowers	electrical	conductivity	to	85%
0·1%	Antimony	,,	,,	,,	,,	90%
0·1%	Arsenic	,,	,,	,,	,,	75%
0·1%	Beryllium	,,	,,	,,	,,	85%
1·0%	Cadmium	,,	,,	,,	,,	90%
0·1%	Calcium	,,	,,	,,	,,	98%
0·1%	Iron	,,	,,	,,	,,	70%
0·1%	Magnesium	,,	,,	,,	,,	94%
0·04%	Manganese	,,	,,	,,	,,	95%
0·1%	Nickel	,,	,,	,,	,,	95%
0·1%	Phosphorus	,,	,,	,,	,,	50%
0·1%	Silicon	,,	,,	,,	,,	65%
1·0%	Silver	,,	,,	,,	,,	97%
1·0%	Tin	,,	,,	,,	,,	55%
1·0%	Zinc	,,	,,	,,	,,	90%

There are many applications which require a relatively high combination of tensile strength or hardness and electrical conductivity. Until recent years, such combinations of properties were not available, and applications requiring high conductivity were met as well as possible by the use of pure copper castings. With the development of the precipitation hardening process, it became possible to obtain high combinations of strength and conductivity together with casting characteristics much superior to those possessed by pure copper. These heat treatments usually involve heating the casting to a temperature in the neighbourhood of $900°C$ for at least one hour, followed by quenching to room temperature and re-heating for one to 5 hours to about $500°C$.

Some High Conductivity Copper Alloys

Silver—Copper (0·03%—0·1% Ag)	Rarely used in cast form. Supplied as wrought material for electrical parts, photogravure plates, etc.
Cadmium—Copper (0·7—1·0% Cd)	Mainly used in form of drawn wire for electrical transmission wires
Chromium—Copper (0·4—0·8% Cr)	Responds to heat-treatment, producing castings of high strength
Tellurium—Copper (0·3—0·7% Te)	Free machining copper alloy for castings or bar stock for repetition machining

Apart from the adverse effects of impurities upon conductivity, soundness and freedom from porosity in the castings must also have serious consideration. Unfortunately, both hydrogen and oxygen have a strong affinity for copper. The former gas dissolves in it, and the latter combines to form cuprous oxide which is soluble in molten copper. When hydrogen and oxygen are both present, gross unsoundness in the form of widespread blowholes is likely to develop from what is commonly referred to as "steam reaction".

As the molten copper cools in the mould its power of holding hydrogen in solution lessens and some hydrogen is expelled. This nascent or atomic hydrogen is very active and at the high temperature involved reacts with any cuprous oxide present to form water vapour. Such an occurrence is often accompanied by visible signs such as the "backing up" of runner and riser heads to form "cauliflowers."

It is apparent, therefore, that during melting, steps must be taken to exclude both hydrogen and oxygen from high conductivity melts. In practice, however, when melting conditions are ideal, it is found that deoxidisation alone is sufficient to prevent the "steam reaction". The need to remove excessive hydrogen from a melt may only arise when the charge consists of cathode copper, which by the nature of its manufacturing process, contains much dissolved hydrogen.

General Melting Procedure

TREATMENT

Melt down under a reducing cover of CUPRIT 8. Deoxidise first with DEOXIDISING TUBES DS containing phosphorus and finally plunge DEOXIDISING TUBES CB containing calcium boride; or DEOXIDISING TUBES L containing lithium.

QUANTITY REQUIRED PER 50 kg MELT
0·5 kg of CUPRIT 8.
DEOXIDISING TUBES DS.
DEOXIDISING TUBES CB or L.
(See sub-section—Deoxidation Techniques.)

RECOMMENDED CASTING TEMPERATURES
Light castings—under 13 mm section. 1250°C.
Medium castings—13-38 mm section. 1200°C.
Heavy castings—over 38 mm section. 1150°C.

DEOXIDATION TECHNIQUES

With few exceptions, the deoxidation of high conductivity copper alloy is best carried out by, first, phosphorus to remove most of the dissolved oxygen and finally by a deoxidant based on calcium

boride or lithium. The latter deoxidants ensure adequate deoxidation while at the same time any small residual quantities will
not adversely reduce conductivity.

METHOD OF DETERMINING QUANTITY OF DEOXIDANT REQUIRED

Owing to the varying degrees of oxidation from one melt to
another the quantities of deoxidant plunged before pouring will vary
accordingly. Since it is only necessary to add sufficient deoxidant to
prevent the "steam reaction," precise recommendations cannot be
given. Instead, a series of simple tests should be made in the foundry
which involve observing the solidification characteristics of a given
melt before any of the castings are poured.

In a standard moulding box, prepare some simple open-topped
cylindrical mould impressions measuring about 75 mm high by
50 mm diameter. When the melt is ready for pouring, but before
deoxidation, a sample of the copper should be ladled into one of the
moulds and the test piece allowed to solidify. If the head rises
appreciably as shown in Fig. 1a, very gassy metal is indicated.
DEOXIDISING TUBES DS containing phosphorus must then be
plunged and further test castings made. At the point when the
quantity of phosphorus added results in a shallow sink in the head
as shown in Fig. 1b, it can be assumed that the residual phosphorus
content of the melt is nil but that approximately 0·008% of oxygen
remains to be removed.

Deoxidation should now be completed by plunging DEOXI-
DISING TUBES CB or L, adding sufficient to produce a test casting
which exhibits a head with a deep sink as shown in Fig. 1c. At this
stage, the melt is in a condition for producing castings free from
porosity.

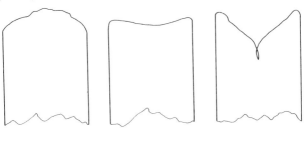

Fig. 1a   Fig. 1b   Fig. 1c

DEGASSING

If a separate degassing treatment using LOGAS 50 is considered necessary, then the briquettes should be plunged into the melt prior to deoxidation and not afterwards.

Degassing by means of LOGAS 50 is not practical for melts of less than 100 kg capacity owing to the fact that melts of small size would drop appreciably in temperature during the degassing process. For degassing a 100 kg melt, one size 1a or one size 1b LOGAS 50 briquette is sufficient.

DEGASSING AND DEOXIDATION TECHNIQUES FOR HIGH CONDUCTIVITY COPPER ALLOYS

Since most high conductivity copper melts are of copper alloy, as distinct from pure copper, it is important to observe degassing and/or deoxidation procedure given in the following notes. This applies particularly when preparing the alloy from virgin materials.

SILVER—COPPER

Silver additions should be made in the form of copper—silver master alloy and introduced into the melt prior to deoxidation but after a degassing operation if the latter is carried out. The dual deoxidation process is recommended, plunging DEOXIDISING TUBES DS followed by DEOXIDISING TUBES CB or L.

CADMIUM—COPPER

Degassing, if necessary, and a full deoxidation treatment by the dual process must be completed before making cadmium additions. Cadmium is often introduced by tapping the molten copper directly on to pure cadmium metal, on transferring the melt from the furnace to a pouring vessel. In crucible melting practice the use of copper-cadmium master alloy may be more convenient and result in lower cadmium losses when added. Ventilation of the foundry should be adequate since molten cadmium evolves toxic brown fumes.

CHROMIUM—COPPER

Chromium additions in the form of copper—chromium master alloy should be made after a degassing operation, if carried out, but before full deoxidation by the dual process. The chromium alloy should be stirred thoroughly into the melt to ensure a homogeneous solution of chromium. A chromium loss varying between 10—30% should be expected depending on the state of oxidation of the melt and alloying techniques. Normally, the chromium additions and a final deoxidation by calcium boride or lithium will deoxidise adequately. Phosphorus additions should only be made if a test casting shows a rising head on solidification. Phosphorus additions must then be limited to that quantity which produces a level sink;

relying upon the chromium additions and CB or L tube to complete deoxidation. An excess of residual phosphorus would up·et response of the alloy to heat treatment.

TELLURIUM—COPPER
    Degas the melt first if this is considered necessary then introduce tellurium in the form of copper-tellurium-master alloy. Finally deoxidise by the dual process using DEOXIDISING TUBES DS followed by DEOXIDISING TUBES CB or L.

COMMERCIAL COPPER CASTINGS

    There is frequently a demand for copper of commercial quality in which a degree of conductivity is sacrificed in the interests of good casting qualities and for ease of machining. These copper alloys may contain tin and/or zinc up to 2%. For these it is possible to adopt a less exacting fluxing and deoxidising routine in comparison to that necessary for a high conductivity copper alloy.

    For commercial quality copper, melting practices as carried out with bronze and gunmetal procedure are usually adopted. That is to say melting is carried out under oxidising conditions and deoxidation performed by means of plunging phosphorus. It is not necessary to employ the Duplex deoxidation technique as previously recommended for high conductivity copper. Therefore, prescribed quantities of phosphorus can be recommended for a given melt and, of course, it is unnecessary to check the degree of deoxidation by means of test castings as previously described.

Melting Procedure

TREATMENT
    Melt down under an oxidising cover of CUPREX blocks. Just before pouring, deoxidise by plunging DEOXIDISING TUBES DS.

QUANTITY REQUIRED PER 50 kg MELT
    0·5 kg CUPREX 1 block.
    Three DEOXIDISING TUBES DS4 (to be plunged in two stages if desired).

RECOMMENDED CASTING TEMPERATURES
    Light castings—under 13 mm section.   1250°C.
    Medium castings—13—38 mm section.   1200°C.
    Heavy castings—over 38 mm sections.   1150°C.

DEGASSING
    Refer to previous remarks in melting procedure for high conductivity copper.

MOULDING PROCEDURE FOR HIGH CONDUCTIVITY AND COMMERCIAL COPPER CASTINGS

General Recommendations

With the exception of chromium copper, copper—nickel—silicon, and beryllium copper alloys, moulding procedure should follow the practice generally adopted in brass and bronze foundries. A mould permeability of at least 25 AFS is desirable in order to avoid blow holes. Green or dry sand moulds may be used but for large castings, over one inch in section, dry sand moulds are preferable.

Where castings are of complex shape or incorporate severe sectional changes, moulds and cores must not be excessively strong because some of the copper alloys are weak at the point of solidification, and tend to crack.

Gating Practice

Every effort should be directed towards obtaining the minimum turbulence on pouring. The illustrations below give some idea of the essential requirements for running and gating high conductivity or commercial copper castings.

SINGLE INGATE — A

The simplest form of gating for small to medium size castings.

1. Tapered sprue to reduce formation of air bubbles.
2. Deep basin to receive first turbulent impact of metal.
3. Ingate tapering out to reduce metal velocity. Note position of ingate at top of sprue basin.

MULTIPLE INGATE — B

For castings of large surface area
where more than one ingate helps
to fill mould uniformly.
1. Sprue and basin as in Example A.
2. Progressively narrowing runner to
   reduce metal velocity ; this assists
   in preventing dross formation.
   Note extension of runner which
   receives first impact of metal
   stream, thus acting as dross trap.

COPPER-ZINC ALLOYS — BRASSES

General Notes

The term "brass" should be applied to alloys of copper in which zinc is the other major element. Casting brasses contain copper between the limits 56—76%. Apart from up to 5% in total of other metals, such as tin, lead, iron, and nickel, the remainder is zinc.

Zinc alloyed with copper, within the range indicated increases the tensile strength and hardness but reduces ductility. This is shown in the following table.

Mechanical Properties of Copper-Zinc Alloys—Sand Cast

Zinc %	Alpha %	Tensile		Elongation % on 2 in. (50 mm.)	Brinell Hardness
		tons/sq. in.	kg./mm.²		
33·7	100	15·0	23·6	63	55
35·0	88	19·1	30·1	59	59
38·0	75	21·5	33·9	56	64
40·1	69	23·0	36·2	55	72
41·8	55	24·6	38·7	50	80
43·7	14	27·8	43·8	27	102
45·5	0	30·4	47·9	27	105

Billets for cold rolling have zinc contents between 27% and 38%. Lower zinc values from 5% to 20%, with balance copper, constitute the alloys known as "gilding metals." In these, other constituents must be at a low level for they are virtually simple alloys of zinc with copper.

Some brass compositions have become known by special names such as:

Dutch metal   78/22 Copper—zinc
Gong metal   76/24 Copper—zinc
Pinchbeck   93·6/6·4 Copper—zinc.
Tobin bronze   60/38/2 Copper—zinc—tin (average composition).
Tombac bronze 80/17/3 Copper—zinc—tin (average composition).

As is seen from the table referred to previously, with zinc at about 34% the brass has a homogeneous structure made up of a single constituent called alpha. By adding more zinc a second constituent begins to appear when prepared specimens are examined under the microscope. It is called beta and increases in quantity as the zinc is raised up to 46%. At this proportion of zinc all the alpha is replaced by beta.

Therefore, brasses are often classified according to their micro-structure. Those with zinc up to 34% are known as alpha brasses, whilst within the range 34—46% zinc are contained the alloys with a duplex structure referred to as alpha—beta brasses. Beyond 46% zinc the alloys again assume a single-phase solid solution structure and form the series of beta brasses. Actually the range of beta brass is very narrow, about 46—51% zinc, so that beyond this limit a duplex structure returns and a new phase makes its appearance, namely gamma.

The gamma solid solution renders the alloy brittle and these compositions have little engineering use except for the 50/50 copper—zinc alloy which is used for brazing.

Melting Technique (Sand Castings)

The brasses can be melted successfully in any of the conventional type furnaces, crucibles accounting for a high proportion of metal melted in foundries all over the world.

Zinc vapour pressure is sufficient to prevent the ingress of hydrogen into the metal and therefore a neutral or even slightly reducing atmosphere is not deleterious. An oxidising atmosphere should be avoided since this would cause heavy melting losses due to the oxidation of zinc.

The molten alloy is at a temperature much above the boiling point of zinc and unless a suitable covering flux is used zinc will be lost both by direct oxidation and also volatilization. Additionally, the flux must be capable of absorbing and holding dirt and oxides without becoming so pasty that it will hold heavy quantities of metal.

As in the case of all metals and alloys, the charge should be melted as rapidly as possible and not be allowed to overheat. "Stewing" the melt will result in increased metal loss and poor quality castings.

Deoxidation, in the normally accepted sense of the word, is not applicable to the brasses. The presence of zinc guards against the oxidation of copper and the oxides of zinc and aluminium (if present) are removed by the relevant fluxes.

Nevertheless, the addition of phosphorus by means of DEOXIDISING TUBES DS is a desirable practice in the case of cast brasses, since it will increase the fluidity of the metal, thus allowing the running of thin sections without unduly superheating the charge.

Procedure

1. Heat up the crucible or furnace.

2. Place in the bottom of the empty pre-heated unit a quantity of CUPRIT 1 blocks equal to 1 kg per 100 kg of the metal to be melted.

3. Commence charging the ingots and scrap and melt down as rapidly as possible, maintaining the surface cover of fused flux intact.
   (NB—Where CUPRIT 49 powder is used, this should be introduced at the same rate as soon as the first part of the charge reaches a pasty condition.)

4. Bring the metal up to pouring temperature and avoid overheating. When the metal is ready for pouring, the moulds should be ready to receive it.

5. Immediately prior to pouring, the metal should be treated with DEOXIDISING TUBES DS at the rate of one DS2 Tube per 50 kg by plunging the requisite number of tubes deep into the heart of the melt and holding immersed for a few seconds.
   (NB—The plunging tool must be pre-heated and should be coated with FIRIT 20 to prevent contamination of the melt and to protect the plunger.)

6. When the metal is at the correct pouring temperature the surface slag should be held back and the metal poured from underneath it. This will curtail "flaring" to the minimum and the unpleasant fuming associated with brass will be almost entirely absent.

7. Recommended pouring temperatures are as follows:

	Under 13 mm section		13—38 mm section		Over 38 mm section	
	(°C)	(°F)	(°C)	(°F)	(°C)	(°F)
Commercial brasses	1150	2102	1100	2012	1070	1958
60/40—65/35 alloys	1100	2012	1050	1922	1020	1868
80/20—70/30 alloys	1150	2102	1100	2012	1070	1958

For charges melted in reverberatory and similar hearth furnaces, CUPRIT 1 blocks are specially recommended. Powder type fluxes are likely to be carried away by the draught and blast from the burners.

Die-castings

The alloys for die-casting are BS 1400:1969, DCB1 and DCB3, both of which contain aluminium. Because of the presence of aluminium a flux capable of dissolving aluminium oxide gives best results and ALBRAL 3 powder is used as follows:

1. Pre-heat the crucible or furnace.

2. Charge in the ingots and scrap and commence melting.

3. When the first part of the charge reaches a pasty stage sprinkle over its surface a quantity of ALBRAL 3 equal to 500 g per 50 kg of metal to be melted.

4. Continue charging and melt as rapidly as possible under the protective cover.

5. When the metal has been brought up to pouring temperature, take a fresh quantity of ALBRAL 3,500 g per 50 kg and, with a perforated saucer plunger, plunge the flux slowly to the bottom of the melt and then with a rotary movement of the plunger "wash" the flux in, bringing it into intimate contact with all parts of the melt in order to cleanse it of alumina particles.
NB—The plunger should be coated and pre-heated as described earlier.

6. After a "washing" period of 2 or 3 minutes, withdraw the plunger and allow the melt to settle for a minute or so.

7. The metal should now be treated with DEOXIDISING TUBES DS at the rate of one DS2 tube per 50 kg by plunging the tube as described previously.

8. As soon as the metal is at correct working temperature — $1100°C$ $(2012°F)$ — it may be ladled out as required, pushing the surface dross aside to leave a clean working area.

9. From time to time, after scrap and ingots have been added to maintain the metal level, the dross should be skimmed away from the surface and fresh ALBRAL 3 should be washed into

the metal as a periodic cleanser. Similarly, DEOXIDISING TUBES DS should be plunged occasionally in order to maintain maximum fluidity. A very considerable improvement in surface finish of gravity diecastings has been reported following the substitution of DS tubes by MG tubes. The addition of a small amount of magnesium to die casting brass apparently improves fluidity and also conditions the surface oxide film to give a much smoother finish to the casting. One MG6 tube should be added to 150 kg approximately of metal, repeating as indicated by results.

Where multiple piece dies are used they are taken apart and dipped in a water suspension of DYCOTE 61 after each pour. In this way the dies are kept cool and the tendency for build up to occur on the face of the dies is reduced. The steel cores are also dipped in the DYCOTE 61 suspension after each cast. Alternative coatings for the die blocks are DYCOTE 36 or 11.

The brass alloy is maintained at 1100°C (2012°F) in the furnace and poured into the dies at about 1050°C (1922°F).

Effect of Added Elements

ALUMINIUM.—Unless added for a definite purpose, as in die-casting brass, it should be absent. Although it gives fluidity and improved definition, it makes founding more difficult in that oxide inclusions and films of alumina tend to cause porosity and un-soundness in castings.

IRON.—In general, small quantities exert a grain refining action. It also has a tendency to increase hardness and tensile strength.

LEAD.—Increases the machinability but must not be allowed to segregate. Lead is insoluble in brass and exists, therefore, as globules. A dispersion of these in as uniform a manner as possible is required.

MANGANESE.—Sometimes used as a deoxidant, its effect is similar to that of iron.

NICKEL.—Improves the mechanical properties and increases resistance to corrosion. Also has a tendency towards grain refinement. Can be introduced as DENSITE 50.

PHOSPHORUS.—Tends to combine with any iron present and increases the hardness. Reduces grain growth and improves fluidity.

SILICON.—Makes founding more difficult but improves corrosion resistance, especially inter-granular attack.

TIN.—Raises the tensile strength and hardness at the expense of ductility but improves corrosion resistance and fluidity.

Other elements sometimes found in small proportions in brass are antimony, arsenic and bismuth. They are regarded as detrimental and should be kept below 0·01% preferably. An exception is made sometimes when arsenic is added up to 0·05% to inhibit dezincification.

Removal of Impurities

1. ALUMINIUM
   In sand cast brass, the major impurity encountered is aluminium and this may be removed quite easily by the use of ELIMINAL. It is well known that in sand castings as little as 0·02% of aluminium is detrimental, since the presence of aluminium oxide films weakens under the structure of the metal, causing porosity and leakage under pressure.

   Aluminium is fairly easily discernible in molten brass and also in the casting. Even in small quantities it forms a scum on the surface of the molten metal, and in the casting its presence is shown by a silky silvery sheen which usually appears on the extremities.

   Up to 0·5% of aluminium may be removed from brass by means of ELIMINAL. If the contamination is heavier than 0·5%, the amount of aluminium should first be reduced by adding metal known to be free from the element.

2. SILICON
   ELIMINAL will also remove silicon from molten brass and the treatment is precisely the same as for the removal of aluminium.

Moulding Practice

The patternmaker's contraction allowance for brass is 1·3% and the liquid shrinkage may be taken as 6—7% by volume.

Brasses may be cast easily in either green sand or dry sand moulds, but the latter is normally used for only the heaviest castings. No difficulty should be experienced in running thin sections, provided the metal is free from non-metallics and that maximum fluidity has been obtained by correct metal treatment.

Normal moulding sand of the natural clay bonded type is quite suitable. Moulds should have good permeability and should be well vented to allow the escape of steam, air, and mould gases. Green, dry, and core sand mixtures suitable for brass moulding are given in Information Sheet No. 57 which deals with bronzes and gunmetals.

Mould Dressings and Sand Additions

Pin-hole porosity is quite prevalent in brass and recent work has shown that this can largely be prevented by the use of suitable mould dressings and by the control of additions to the facing sand. This pin-hole porosity is generally revealed when the surface of the casting is polished prior to plating operations, and it occurs particularly in the sections which require the higher pouring temperatures.

Wide use has been made of wheat flour and wheat flour substitutes but, although these materials provide a smooth surface on the casting, they encourage sub-surface pin-hole porosity due to a metal-mould reaction which occurs with the wheat flour.

0·5% pelletted pitch or 2% TERRADUST 2 added to the facing sand gives an excellent surface to the casting without the danger of pinhole porosity.

Self-drying or fired dressings on the surface of the mould will also produce an excellent finish free from pin-holes and the use of either MOLDCOTE 6, MOLDCOTE 9, or MOLDCOTE 10 is recommended, particularly for green sand and silicate bonded moulds.

Running, Gating, and Feeding

The running of brass castings does not present any real problem. Excessive turbulence in the mould should be avoided, but there is no need for elaborate runner systems as in the case of high tensile brass.

Generous feeder heads should be provided over heavy sections and the efficiency of these feeder heads may be greatly improved by the use of FEEDEX, FEEDOL, and KALMIN 44.

Recommended Sand Cast Test Bar Designs for Brass Alloys

Recommended sand cast test bar designs for brass alloys

B.N.F.M.R.A. 'cast to shape' test bar pattern dimensions
(not to scale)

The length L may be varied to suit the testing machine

All dimensions in millimetres
Sand cast test bar

The length L may be varied to suit the testing machine

All dimensions in millimetres
Chill cast test bar

COPPER ALLOYS—BRONZES & GUNMETALS

General Notes

The original bronze alloys were of copper with tin as the main alloying element. Such a wide variety of bronze and gunmetal alloys now exist, however, that it has become necessary to classify them according to their basic chemical composition. The following examples are given and on pages 179-183 some current British Specifications can be seen.

TIN BRONZES

Copper based alloys having tin as the main alloying element. Alloys having between 3% and 7% tin are not normally produced as castings but in billet form for subsequent drawing or rolling operations. Alloys containing between 15% and 20% tin are widely used for the casting of bells. As the tin content reaches this magnitude, the alloy becomes increasingly harder and difficult to machine.

PHOSPHOR BRONZES

Copper based, having from 5% up to 10% tin as the main alloying constituent together with an intentional phosphorus addition which may vary from 0·5 to 1·0%. These alloys are used to make what are widely known as phosphor-bronze bearings. Tin provides the necessary toughness for the alloy whilst phosphorus additions give castings the necessary hardness and therefore, wear resistance.

GUNMETAL

Copper based, with tin and zinc as the alloying constituents. Tin usually predominates and an example of this type of alloy is Admiralty gunmetal, 88% copper, 10% tin, 2% zinc. This alloy is used for a variety of purposes, ranging from casting of gears and bearings to hydraulic castings and castings subjected to moderate stresses.

LEADED GUNMETAL

Copper based with tin, zinc, and lead present in varying proportions. Numerous alloys of this type exist, a widely known example being 85/5/5/5 alloy, in which tin, zinc and lead are present in equal proportions. These alloys are extremely popular throughout the world and are used for a number of purposes, mainly in the casting of hydraulic components, plumbers' fittings, etc. They are also used as bearing alloys operating under moderate loads.

LEADED BRONZE

Copper-tin, or copper-tin-phosphorus alloys containing from 5-15% lead. These alloys are used almost exclusively for casting heavy duty bearings, of which wear resistance combined with the ability to undergo plastic deformation is required.

Bronzes and Gunmetals—Melting Practice

Effect of Gas Absorption

Bronze and gunmetal alloys are usually melted in coke, gas or oil fired crucible furnaces. Many foundries also employ small oil or gas fired reverberatory furnaces and, in some instances, electric melting by means of an arc or induction furnace is employed.

In the molten condition, these alloys may come into contact with gases which include oxygen, nitrogen, water vapour, hydrogen, carbon monoxide, carbon dioxide, and sulphur dioxide. Carbon dioxide is generally considered to be inert to molten bronze but carbon monoxide and sulphur dioxide have often been reported as minor sources of porosity; the effect of sulphur as metallic sulphides is slight.

Hydrogen is the chief source of trouble for the bronze or gunmetal founder and this may be derived from the products of incomplete combustion during melting, or from water vapour in the atmosphere. Hydrogen is not so soluble in copper alloys as in pure copper, but this gas can cause severe porosity when the gas is ejected during solidification of a casting.

The Steam Reaction

Water vapour reacting with molten copper forms cuprous oxide and hydrogen, both of which are dissolved in the melt. During solidification hydrogen is ejected and reacts immediately with cuprous oxide to form steam. During solidification of a casting the steam evolved causes severe porosity. It is suggested that this reaction is the main cause of porosity in bronze, except that the reacting oxide is oxide of tin.

If this is so then by eliminating both oxygen and hydrogen from a melt the formation of porosity should be prevented. For this reason, it is often advisable to both degas and deoxidise a bronze or gunmetal melt before casting.

The Oxidation-Deoxidation Melting Process

If during melting oxidising conditions are maintained, this reduces pick-up of hydrogen gas to a minimum because relatively large volumes of both hydrogen and oxygen cannot exist together as shown in the graph (over page). By maintaining a slightly oxidised melt, therefore, hydrogen pick-up can be prevented and it is only necessary to deoxidise a melt before pouring to ensure freedom from both hydrogen and metallic oxides. This method of melting is known as the oxidation-deoxidation process.

Oxidising conditions are achieved in two ways: by employing an oxidising covering flux such as CUPREX and by adjusting the

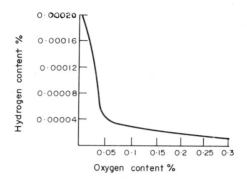

furnace atmosphere to slightly oxidising burning conditions. Further details are given below.

Degassing

The degassing of bronze or gunmetal melts by means of bubbling inert gases through the melt is widely practised and nitrogen is often used for the purpose. There are disadvantages with nitrogen, however, particularly with respect to the difficulty of handling gas cylinders in a production foundry. Metal tubes for injecting the gas into the metal cannot be used and even the best non-metallic tubes are fragile. Again, in order to overcome the resistance of pressure in deep ladles or crucibles, a fast flow of nitrogen is required. It is generally assumed that nitrogen delivered from a cylinder is dry, provided that pressure does not fall below 20 atmospheres. However, moisture determinations on commercial nitrogen have shown that the moisture in a full cylinder was 0·32 milligrams per litre and 2·7 in a cylinder three-quarters empty. Therefore, 50 litres of nitrogen will contain 80 ml of water in vapour form at a temperature of 1100°C for a full cylinder and approximately 650 ml for a three-quarters empty cylinder.

Alternative Degassing Methods

Alternatively LOGAS 50 briquettes may be plunged into the melt using a suitably protected plunging tool. These briquettes evolve inert gases at the temperatures involved, and degassing—as with nitrogen—is accomplished by a process of diffusion into the rising bubbles. This system is more robust than degassing with nitrogen gas, has none of the limitations, and is equally effective.

A degassing method based on scavenging chemicals is also available. A powdered product known as REGENEX is plunged and well "scrubbed" into the metal with a protected bell-plunger. This method is quick and convenient but may not be the most suitable where very low residual gas levels are required or where close chemical specifications must be held.

Melting Technique—Sand Castings

1. Preheat the crucible or furnace.
2. Place in the bottom of the hot, empty crucible sufficient CUPREX 1 blocks using 1 kg per 100 kg or 1% of the charge weight.
3. Charge in the ingot metal and/or scrap. Alternatively, CUPREX 100 powder can be added with the charge, or at an early stage of melting, again using the 1% ratio.
4. Melt and bring to pouring temperature as rapidly as possible. During melting CUPREX evolves oxidising and scavenging gases which combine with hydrogen and remove it from the melt. An oxidising flux cover is also formed which protects the melt from further hydrogen absorption.
5. Immediately before casting, deoxidise the melt thoroughly by plunging DEOXIDISING TUBES DS using one DS3 tube per 50 kg of melt. The object of this is to introduce enough phosphorus to give a residual phosphorus content of 0·02% which signifies adequate deoxidation and maximum fluidity.
6. If the charge materials contain scrap which is very oily or dirty a large quantity of hydrogen will find its way into the melt. Thus extra degassing may be necessary in such circumstances; also in the case of special castings required to withstand relatively high internal pressure or to be specially sound and free from porosity. Then, before using the DEOXIDISING TUBES, plunge LOGAS 50 briquettes or REGENEX 6 using 100 g to 250 g per 50 kg of melt.
7. Check the correct pouring temperature as given in the accompanying table, skim and cast without delay, taking care to prevent slag entering the mould cavity.
8. Molten bronze should never be held in the furnace waiting for moulds to be ready, the moulds must be prepared in advance to receive the metal as soon as its melting and fluxing treatment is completed.

Reverberatory Furnace Melting

CUPREX 1 blocks are especially suited for fluxing gunmetal and tin-bronze charges melted in open flame reverberatory type furnaces such as the Sklenar. One obvious advantage of the flux in block form is that, unlike powder flux, it is not carried away by the hot air currents and blast pressure from the oil burners.

The CUPREX blocks are placed on the hearth of the furnace in a ratio of 1% of the charge weight and the ingot metal and scrap charged on top of them. Their action is to evolve oxidising and scavenging gases which rise through the melt, cleaning it and removing hydrogen. As already indicated, in the final stages they melt to form a protecting slag cover which guards the molten charge from hydrogen pick-up from the furnace atmosphere. This is an important function considering the relatively large surface area of melt exposed to the furnace gases in shallow hearth furnaces.

Recommended Casting Temperatures

Copper-tin-zinc-lead	Under 13 mm section (°C)	13-38 mm section (°C)	Over 38 mm section (°C)
83/3/9/5	1180	1140	1100
85/5/5/5	1200	1150	1120
86/7/5/2	1200	1160	1120
88/10/2	1200	1170	1130
90/10/(phosphor bronze)	1120	1100	1030

Metal-Mould Reaction

Molten metal poured into a sand mould is immediately confronted with a steam atmosphere which is generally oxidising to the metal. In this way the oxygen becomes fixed and the oxide skin formed is, in most cases, tough and impermeable, thereby stifling further reaction. However, with certain alloys the skin of oxide does not prevent further reaction.

Occurrence of metal-mould reaction depends on the presence of highly reactive elements, the most important of which are magnesium and phosphorus. It occurs also in both aluminium and copper-base alloys which contain small amounts of boron. Substantial metal-mould reaction is produced also in copper-base alloys which contain lead and silicon present together.

The water vapour formed when the molten metal enters the mould, having given up its oxygen to form an oxide skin, releases a quantity of hydrogen. Such "newly born" hydrogen is very active and can enter into solution with the molten metal unless the oxide skin formed on it is very strong and protective.

In copper-base alloys which contain phosphorus the reaction does not form an oxide but probably a phosphate skin which remains liquid in temperatures near the solidification point of the alloy. This liquid does not provide a sufficient barrier on the alloy surface to stifle the reaction. Similarly, in phosphate-free copper-base alloys which contain lead and silicon, the oxidation of these two elements at the metal-mould interface produces lead silicate which acts similarly to phosphate.

With lead or lead-free gunmetal containing zinc, a phosphorus content of 0·03% or above is enough to cause the reaction. A residual phosphorus content of 0·06%-0·08% is usually sufficient to produce an appreciable effect. In general, the degree of gas absorption increases as the pouring temperature is raised and vice versa. For gunmetal alloys and bronzes the reaction can be prevented by coating the moulds and cores with MOLDCOTE 13 which contains a spirit suspension of aluminium.

Metal-mould reaction can sometimes be used to advantage for certain pressure-resisting bronze-castings such as taps and valve bodies. The idea is to impart a controlled gassing which serves to disperse concentrated liquid shrinkage and distribute it as uniform microporosity throughout the mass of the casting. To accomplish this effectively the alloy must be gas free as cast and contain sufficient phosphorus to set up a limited metal-mould reaction—this is dependent on the casting section.

Pin-holes

The occurrence of these on the surface of bronze castings which are required to be bright plated is a not uncommon source of trouble. Investigations carried out by the BNFMRA have shown the cause as a form of metal-mould reaction due to the presence of wheat flour used as an additive to the facing sand or as a mould dressing.

Whilst the defect occurs on thin leaded gunmetal castings, made in such moulds and cast at temperatures above 1160°C, it is much less prevalent in lead-free alloys. Phosphorus contents of the order of 0·08% cause an increase in the amount of pin-holing.

Pin-holing can be reduced by omitting cereal flours from the sand or mould facing and is prevented by spraying with MOLDCOTE 13. These measures may produce castings with a rougher surface than when using flour but this may be improved without risk of pin-holing by adding 0·5% or more of pelletted pitch to the facing sand.

Gravity Die-castings

In melting down the initial bale-out charge, CUPREX 1 blocks should be employed on a basis of 1% of the charge weight. Thereafter melting will be continuous with fresh ingot metal and returned scrap being added from time to time to maintain the bath of molten alloy. REGENEX should be plunged at appropriate intervals to keep the liquid metal clean and of low gas content.

Following the degassing operation, strong deoxidation is carried out by plunging one DEOXIDISING TUBE DS4 to each 25 kg of melt. This also has the effect of promoting fluidity to the maximum degree. Pouring temperatures range from $1150°$ to $1200°C$ and the die should be of heat-resisting cast-iron or steel to stand up successfully.

DYCOTE 36 is recommended for spraying the die blocks, which should be sunk with as much taper as dimensional tolerance will allow, otherwise extraction difficulties will become a major problem.

Where dies are dissembled after each cast and are quenched into a dispersion of lubricant in water DYCOTE 61, 11, or 38 are the products to use. DYCOTE 11 and 38 are aqueous dispersions of graphite and should be diluted before use. DYCOTE 61 is similar but contains additions of chemicals designed to inhibit zinc oxide build-up on the die face. With all three products, dilution rates of the order of 1 part product to 15 parts of water are common, while for simple castings the proportion can often be increased to 1:30 or even more.

Eliminate any tendency to build up back pressure by cutting venting grooves in the die at as many points as design permits.

Before casting commences, always pre-heat the dies to a working temperature of $350°$-$400°C$ with the idea of limiting thermal shock and blowing.

Removal of Aluminium

Aluminium is a common and very deleterious impurity in sand cast gunmetals and bronzes. As little as 0·01% is enough to cause leakage of pressure type castings. This is because the aluminium oxidises and forms tough films and "stringers" which become trapped in the solidifying castings. The presence of the oxide films also interferes with self-feeding properties.

Fortunately, it is not unduly difficult to remove aluminium in quantities up to about 0·5% from molten bronze and gunmetal by means of ELIMINAL.

Founding Properties

Where maximum soundness and pressure-resisting properties are called for the leaded gunmetals are to be preferred. Compositions such as 86/7/5/2 and 85/5/5/5 are best for the production of sound and strong pressure castings for use at atmospheric temperatures. They give maximum pressure tightness coupled with ease of handling in the foundry.

Nickel Additions

Part of the tin content of these alloys can be replaced by nickel when the mechanical properties and corrosion resistance are improved without sacrifice of any of the good founding qualities.

Nickel additions of 1 to 3·5% to pressure bronze castings add measurably to density, tensile yield strength and toughness, especially for tin contents of 2% or more. In addition to grain refining and densening, nickel bolsters the strength of compositions in the low tin range and materially improves their performance at elevated temperatures. When nickel additions to bronzes exceed 4% the alloys are susceptible to age-hardening.

If a nickel addition is made to an alloy, the fluidity at any given temperature is affected and a suitable increase in the pouring temperature should be made. An easy and accurate method of making the addition is by means of DENSITE.

Running and Feeding

The literature on running and feeding bronzes and gunmetals is extensive. In general those methods best suited to long freezing range alloys should be used, and the most commonly used systems appear to be the non- or only slightly pressurised systems based on a 1:4·1 type of cross-sectional area ratio. Besides giving a good skimming effect, this type of downsprue/runner bar/ingate system can provide a worthwhile source of feed metal to the casting as long as the gate remains unfrozen. Indeed, many thousands of castings (shell mouldings in particular) such as taps, valves, cocks, etc., are made in this way without any supplementary form of feeding.

Where additional feed to that obtained from the running system as outlined above is required, quite generous risers should be provided. Because of the long freezing range and pasty method of solidification, risers generally will have to be of greater section than the part of the casting they are feeding. Again, considerable economies can be obtained by utilising FEEDEX exothermic or

KALMIN 44 insulating riser sleeves. These will effectively reduce heat loss through the walls of the riser, but having closed off that route it becomes increasingly important that loss to atmosphere is controlled as much as possible. All risers should therefore be covered with a layer (about 13 mm thick) of FEEDOL 16 or 17 immediately on completion of casting. It is better if the risers are short-poured slightly so that the optimum effect from the anti-piping compound is obtained.

MOULDING SAND MIXTURES
FOR COPPER-BASE ALLOYS

A variety of sand mixtures are used, both naturally and synthetically bonded with mineral clay such as bentonite. In Great Britain the naturally occurring clay bonded red sands are common and some facing sand mixtures recommended by R. F. Hudson are given below.

Green Sand Mixture

10% New Red Sand.
90% Old Floor Sand (comprising old red sand).
Add approximately 6% of water and mill for 5 minutes. If desired, 0·5% of pelleted pitch can be added to provide an improved "as-cast" surface to the casting.

Dry Sand Mixture

12% New Red Sand.
88% Old Red Sand.
Add 8% of water, mill and finally aerate.

Properties of a Typical Natural Moulding Sand

Moisture content	5·0-7.0%.
Permeability No. (AFS)	25-30%.
Compression strength (AFS)	4·5-6·0 lb per sq. in.
Clay content (below 20 microns)	10·0-12·0%.
Ignition loss on dried sample (volatile matter content)	Normally less than 0·5%
Flowability (AFS)	65%.

SIEVE ANALYSIS: BRITISH STANDARD SIEVE NOS.

Retained on BSS 8-mesh sieve	Nil
Retained on BSS 10-mesh sieve	0·1%
Retained on BSS 16-mesh sieve	0·3%
Retained on BSS 22-mesh sieve	0·6%
Retained on BSS 30-mesh sieve	0·8%
Retained on BSS 44-mesh sieve	1·2%
Retained on BSS 60-mesh sieve	4·3%
Retained on BSS 100-mesh sieve	18.6%
Retained on BSS 150-mesh sieve	32·9%
Retained on BSS 200-mesh sieve	15·6%
Through 200-mesh sieve	14·4%
Clay grade	11·2%

Loam Mixture

44% Old Red Sand.
44% New Red Sand.
7·5% Horse Manure.
3·5% Sawdust.
1·0% Cowhair.
Add 20% water and mix thoroughly.

Core Sand Mixtures

A suitable mixture based on natural sand and bentonite mineral clay is as follows :

80% Old Red Sand.
18% New Red Sand.
2% Bentonite.
Add 7½ to 8% of water and mill thoroughly.

Oil Sand Cores

Core sand mixtures bonded by linseed oil and cereals, or blended semi-solid oil type binders are in widespread use and a number of suitable mixtures can be found on page 88. See also the section (page 93) concerned exclusively with core binders.

CASTING OF CORED AND SOLID
BRONZE STICKS

The first part of these notes is concerned with the casting of sticks up to 1 m in length and from 25 to 150 mm in diameter. This type of casting formerly formed a large part of the bronze foundry industry since it provided the starting point of the large extrusion and rolling sectors. While some cored stick is still made by these methods, the process has been very largely superseded by continuous and semi-continuous casting methods. These newer techniques demand much more capital outlay and technical process control, but the end product is far more consistent and reliable.

Moulds

Split moulds with spigot-joints, or one piece cannon moulds are both in general use, but the former type are preferred owing to the ease with which their surfaces can be cleaned and dressed. Moreover, unlike cannon moulds, which have to be tapered in order to make withdrawal of the casting easy, a split mould produces a parallel casting. A suitable dressing for applying to a warm mould is INGOTOL 16, but for the best results, particularly in the case of high phosphorus alloy, MOLDCOTE 13 applied in a similar way is recommended.

For moulds which will have a long life, high resistance to repeated casting and, hence, freedom from premature crazing, Hudson* suggests the following grey iron composition :

Total carbon ...	...	...	...	...	3·0—3·3%
Silicon	...	...	...	...	1·2—1·6%
Manganese	...	...	...	...	0·8—1·2%
Sulphur	...	...	...	...	0·12% max.
Phosphorus	...	...	...	...	0·3% max.

The soundness of a solid or cored chill casting depends largely on obtaining the ideal balance between pouring temperature and rate of pouring. Ideally, these factors should be so adjusted that as the mould is filling, solidification takes place in one direction; from the base of the mould upwards and not by process of simultaneous solidification in all parts of the casting. This demands relatively slow pouring rates which are achieved by allowing the melt to fall into the mould by way of a dozzle box in which a hole is drilled in the base.

Generally pouring rates are best found by practical experiment, but in the following table are given suggested pouring hole diameters suitable for various casting diameters.

*Hudson R. F. Non-ferrous Castings.—Chapman & Hall.

Description	Aperture diameter
1 in (2·54 cm) sticks	3/32 in (2·4 mm)
2 in (5·00 cm) sticks	5/32 in (4·0 mm)
3 in (7·62 cm) billets	3/16 in (4·8 mm)
4 in (10·2 cm) billets	7/32 in (5·6 mm)
6 in (15·2 cm) billets	1/4 in (6·4 mm)

Cores for Hollow Stick Castings

Cores may be machined from mild steel but they must have a taper of at least 15 mm per metre length of the core. Various methods of coating the core surface are in use, to prevent sticking of the core to the casting and to make withdrawal of the core easy. One method is to strickle a coating od dry sand mixture around the core or wrap two layers of thin asbestos paper around it. In both cases, however, the core must be returned to a drying oven for either hardening of the sand coating or to drive off uncombined moisture from the asbestos paper. Following this a dressing is advised and MOLDCOTE 13 is particularly recommended.

Continuous and Semi-continuous Casting Methods

As indicated, these high production methods are rapidly superseding the older labour intensive craft methods where the demand can justify the considerable capital outlay involved.

In theory the processes are similar and quite simple. A bulk melting furnace(s) feeds molten metal into a separately heated holding furnace. From the base (semi-continuous casting) or wall (continuous casting) of the holding furnace, metal is fed out via a water-cooled copper and graphite die and withdrawn slowly and continuously as solid rod, strip, etc. In semi-continuous casting one or more billets, etc. (depending on the number of dies), are cast vertically, the limiting factors being working room under the die and furnace capacity; the process is usually operated on a batch basis. The continuous system, once in operation, is limited only by the supply of molten metal and the ability to handle the sections produced. A simplified schematic layout of a typical system is shown below.

In practice a large number of factors must be carefully balanced and stringently controlled in order that the processes can function at all.

'Schematic diagram of a continuous casting unit'

A delicate temperature balance between the various parts is clearly essential. The only point at which metal treatment as such is possible or desirable is in the bulk melting unit, so it is important that correct melting, fluxing, and deoxidation techniques are used as a matter of course.

Suitable Test Bars for Use with Bronzes and Gunmetals

BNFMRA "Cast to shape" test bar pattern dimensions (not to scale).

This test bar shall be used for the following alloys: LG1, LG2, LG4, G1.

This test bar may also be chosen when required for the following alloys: PB4, LPB1, LB2, LB4, PB1, PB2, CT1, LB5, LB1, G3.

TEST BARS SUITABLE FOR USE WITH BRONZES AND GUNMETALS

DTD PATTERN BAR

Gated here and two bars
per mould poured from
common runner

CROWN PATTERN BAR

THE RECOVERY OF NON—FERROUS SWARFS, SCRAPS, ETC.

There is a very considerable arising of finely divided swarf, turnings, etc., from all sectors of the non-ferrous metal industry, and it is important that as much metal as possible is recovered for re-use. A similar reclamation technique will be used whatever metal or alloy is being melted, but different flux compositions will be required as indicated by the various metal characteristics.

Furnace Types

The furnaces used are generally of the slowly rotating cylindrical reverberatory type holding 5 tonnes or more of metal when full. Induction furnaces are, however, being used more and more, and are particularly suitable for very finely divided scraps since there is a relatively static atmosphere immediately in contact with the metal. At the same time a good "boil" can ensure scrap is melted and assimilated quickly, thus keeping metal losses at a minimum.

Procedure

Whichever alloy or mixture of scraps is being recovered, the operating sequence is similar for both of the furnace types mentioned. Some solid or heavy scrap is first melted to produce a "heel" of metal in the furnace. This is then provided with a fluid cover of RECUPEX, and further quantities of the flux are scattered over the swarf, etc., waiting to be charged. Depending on the fineness of the scraps and their cleanliness, the amount of flux required can vary from around 2% by weight of the charge up to 5%. Some fluxes can attack refractories, etc., if used in excessive amounts, so it is always best to obtain advice beforehand. Charging can then proceed, the scrap being added to the furnace at approximately the same rate at which it is melting so that wide temperature fluctuations in the pool of molten metal do not occur. The amount of flux used should be just sufficient to maintain a slag cover of the requisite fluidity in order that the melt is protected at all times.

Contamination by Aluminium

After recovering incompletely segregated or mixed lots of scrap it is quite common that the resulting metal will be contaminated with aluminium. Unless the presence of this element is required in the alloy being produced it is usually regarded as a harmful contaminant, particularly if the metal is to be used for sand castings.

Before casting, therefore, and provided the amount of aluminium present does not exceed approximately 0·5% the melt should be treated with ELIMINAL. Above 0·5% contamination it is preferable to dilute with aluminium-free metal until that level is reached or the amount of dross produced can become unwieldly. The temperature of the metal is raised to around 1150°C and the RECUPEX slag is drained off or removed into a slag pot. ELIMINAL is then applied to the bath of metal and brought into as intimate contact as possible by rabbling and stirring. The amount of aluminium remover required will depend on the degree of contamination but will generally be about four times the weight of aluminium to be removed. The treatment is best applied in stages using part of the ELIMINAL each time. Allow the flux to fuse on the surface before working it into the melt.

Choice of Materials

Alloy	Flux cover	Deoxidant
Brass 60/40, 65/35, 70/30	RECUPEX 112, 119	—
Bronzes and gunmetals	RECUPEX 118, 119 or 55	DS Tubes
HT Brass and Al bronze	ALBRAL 16	E Tubes
Nickel and cupro nickel	RECUPEX 118, 119 or 55	MG(R) Tubes

PICKLING OF BRASS AND GUNMETAL
CASTINGS TO IMPROVE APPEARANCE

There are various opinions among buyers as to the colour of brass and gunmetal castings which they purchase. It is important that the castings should have a good appearance and this may be achieved as follows :—

Make up a solution containing :—

66% concentrated nitric acid,

17% water,

17% concentrated sulphuric acid.

The water should first be placed into a suitable vessel (stone jar). Then gradually add the sulphuric acid with constant stirring. Cool the solution, and slowly introduce the nitric acid, stirring the while. Cover the mixture with a suitable lid in order to prevent evaporation (metal vessels or covers cannot be used for this solution).

A wide-necked vessel is desirable, and side by side with it, should be a similar vessel containing clean water. Castings are fastened on to a wire and quickly dipped, first in the acid and then in the water, being well swilled till free from acid.

The castings, after drying in sawdust and polishing, have a pleasing appearance. It is important that the water is clean and changed often, because the casting must be free from acid after washing.

RECOMMENDED STANDARD MELTING AND FLUXING PROCEDURE FOR NON-FERROUS METALS AND ALLOYS

COPPER — HIGH CONDUCTIVITY

Copper of over 99% purity. Can be produced with electrical conductivity in excess of 85% I.A.C.S.

Treatment

Melt down under a reducing cover of CUPRIT 8. Deoxidise by plunging, at least twice, DEOXIDISING TUBES CB (containing calcium boride) or DEOXIDISING TUBES L (containing lithium) immediately before casting. (See Information Sheet No. 43.)

Quantity Required per 50 kg Melt

250-500 g CUPRIT 8.
1 DEOXIDISING TUBE CB5 (to be plunged),
or 1 DEOXIDISING TUBE L1 (to be plunged).

Effect

Reducing atmosphere combined with a cleansing slag. Strong deoxidation without deterioration of conductivity properties.

Recommended Casting Temperatures

Light castings, under 13 mm section	1250°C.
Medium castings, 13-38 mm section	1200°C
Heavy castings, over 38 mm section	1150°C.

For the most effective removal of gas plunge LOGAS 50 briquettes.

BRASSES — COMMERCIAL

Non-specification. These alloys usually contain a small percentage of aluminium as an impurity

Treatment
Melt down under a fluid cover of ALBRAL 2 and plunge a further quantity a few minutes before pouring. Alternatively, if aluminium is to be removed, use ELIMINAL 2. In either case finally plunge DEOXIDISING TUBES DS.

Quantity Required per 50 kg
0·5 kg ALBRAL 2 (¾ as surface cover, ¼ to be plunged), or 0·5 kg ELIMINAL 2 (used as above).
1 DEOXIDISING TUBE DS2.

Effect
ALBRAL reduces oxide (alumina) formation while ELIMINAL removes aluminium, both giving a general cleansing action. Increased fluidity obtained by "deoxidation" treatment.

Recommended Casting Temperatures
Light castings, under 13 mm section	1150°C
Medium castings, 13-38 mm section	1100°C
Heavy castings, over 38 mm section	1070°C

BRASS ALLOYS (SAND CAST)

For example, 60/40 and 65/35 alloys containing no aluminium. For such applications as pipe-line and boiler fittings, etc.

Treatment
Melt down under a fluid cover of CUPRIT 49 or use CUPRIT 1 blocks. Finally plunge DEOXIDISING TUBES DS immediately before casting.

Quantity Required per 50 kg Melt
0·5 kg CUPRIT 49 (as surface cover), or 1 x 0·5 kg CUPRIT 1 block (at bottom of crucible).
1 DEOXIDISING TUBE DS2.

Effect
Fluid slag curtails zinc loss and has a general cleansing action. Increased fluidity obtained by "deoxidation" treatment.

Recommended Casting Temperatures

Light castings, under 13 mm section	1100°C
Medium castings, 13—38 mm section	1050°C
Heavy castings, over 38 mm section	1020°C

BRASS ALLOYS (DIE CAST)

For example, 60/40 alloy and silicon brass. Small percentages of aluminium (up to 0.5%) and silicon are usually present.

Treatment

Melt down under a non-fluid cover of ALBRAL 3 and plunge further small quantities as furnace is recharged from time to time. Dry dross can either be pushed aside during ladling out or skimmed off. Occasionally plunge DEOXIDISING TUBES DS.

Quantity Required per 50 kg Melt

0.5 kg ALBRAL 3 (as initial cover).
250 g ALBRAL 3 (to be added at pouring temperature).
1 DEOXIDISING TUBE DS2.

Effect

Oxide formation and zinc fuming curtailed together with an overall cleansing action. Increased fluidity obtained by "deoxidation" treatment.

Recommended Casting Temperatures

Light castings, under 13 mm section	1100°C
Medium castings, 13—38 mm section	1100°C
Heavy castings, over 38 mm section	1100°C

BRAZING METAL

For example, 80/20 alloy.

Treatment

Melt down under a fluid cover of CUPRIT 49 or use CUPRIT I Blocks. Finally plunge DEOXIDISING TUBES DS immediately before casting.

Quantity Required per 50 kg Melt

0.5 kg CUPRIT 49 (as surface cover)
or 1 x 0.5 kg CUPRIT I block (in bottom of crucible).
1 DEOXIDISING TUBE DS2.

Effect

Maximum cleansing action with curtailment of zinc loss obtained by fluid slag. Increased fluidity obtained by "deoxidation" treatment.

Light castings, under 13 mm section	1150°C
Medium castings, 13-38 mm section	1100°C
Heavy castings, over 38 mm section	1070°C

GUNMETAL (TIN BRONZE) ALLOYS

For example BS 1400-G1, i.e. 88/10/2 (Admiralty).
BS 1400-LG2, i.e. 85/5/5/5.
BS 1400-LG3, i.e. 86/7/5/2.

Treatment

Melt charge on top of CUPREX 1 block in bottom of crucible or melt down under a cover of CUPREX 100. Deoxidise by plunging DEOXIDISING TUBES DS immediately before casting.

Quantity Required per 50 kg Melt

1 x 0·5 kg CUPREX block 1 (in bottom of crucible),
or 0·5 kg CUPREX 100 (as surface cover).
1 DEOXIDISING TUBE DS3.

Effect

The classical "oxidation-deoxidation" technique, i.e. oxidising and scavenging action, prevents hydrogen pick-up and cleans the melt, while deoxidation removes excess oxygen, reduces oxides and improves fluidity. (Residual phosphorus not less than 0·02%.)

Recommended Casting Temperatures

Light castings, under 13 mm section	1200°C
Medium castings, 13-38 mm section	1160°C
Heavy castings, over 38 mm section	1120°C

For the most effective gas removal plunge LOGAS 50 briquettes.

PHOSPHOR BRONZE

For example, BS 1400-PB3. For general sand castings.
BS 1400-PB1. For bearings.
BS 1400-PB2. For gear blanks and bearings (good rigidity).

Treatment

Melt charge on top of CUPREX 14 blocks in bottom of crucible or melt down under a cover of CUPREX 140. Plunging of DEOXIDISING TUBES not necessary unless it is intended to increase phosphorus content.

Quantity Required per 50 kg Melt

1 x 0·5 kg CUPREX 14 block (in bottom of crucible),
or 750 g CUPREX 140 (as surface cover).

Add one DEOXIDISING TUBE DS4 to compensate for loss of phosphorus during melting.

Effect

Scavenging action removes gas and has general cleansing effect. Tin "sweat" curtailed. DEOXIDISING TUBES DS increase phosphorus content where necessary.

Recommended Casting Temperatures

Light castings, under 13 mm section	1100°C
Medium castings, 13-38 mm section	1070°C
Heavy castings, over 38 mm section	1040°C

For the most effective gas removal plunge LOGAS 50 briquettes.

SILICON BRONZE

For example, DTD 355, Everdur and PMG metal.

Silicon	...	...	...	...	...	1·5—5%
Manganese	...	...	...	...	...	1·5% max.
Iron	...	...	...	...	...	2·5% max.
Zinc	...	...	...	...	...	5·0% max.
Copper	...	...	...	...	...	balance.

Treatment

Melt down under a fluid cover of ALBRAL 2 and when pouring temperature is reached plunge in further small quantities. Finally plunge DEOXIDISING TUBES E immediately before casting.

Quantity Required per 50 kg Melt

0·5 kg ALBRAL 2 (¾ as a cover, ¼ to be plunged).
1 DEOXIDISING TUBE E3.

Effect

Curtails oxidation and reduces metallic oxides combined with general cleansing action. "Deoxidation" treatment coalesces suspended non-metallics and improves fluidity.

Recommended Casting Temperatures

Light castings, under 13 mm section	1200°C
Medium castings, 13-38 mm section	1160°C
Heavy castings, over 38 mm section	1120°C

For the most effective gas removal plunge LOGAS 50 briquettes.

MANGANESE BRONZE (High Tensile Brass)

For example BS 1400-HTB1.

Manganese	...	...	...	...	...	3% max.
Aluminium	...	...	...	...	...	2·5% max.
Nickel	...	...	...	...	...	1% max.
Iron	...	...	...	...	...	0·5-2%
Tin	...	...	...	...	...	1·5% max.
Lead	...	...	...	...	...	0·5% max.
Zinc	...	...	...	...	...	balance.
Copper	...	...	...	...	...	55-63%

Treatment

As for silicon bronze (above) but finally plunge two DEOXIDISING TUBES E1 per 50 kg melt.

Recommended Casting Temperatures

Light castings, under 13 mm section	1080°C
Medium castings, 13-38 mm section	1040°C
Heavy castings, over 38 mm section	1000°C

For the most effective gas removal plunge LOGAS 50 briquettes.

ALUMINIUM BRONZE

For example BS 1400-AB1 DTD 174A. Suitable for both sand and die casting.

Aluminium	...	...	...	...	...	9·5%
Nickel	...	...	...	...	...	1·0% max.
Manganese	...	...	...	...	...	1·0% max.
Iron	...	...	...	...	...	2·5%
Copper	...	...	...	...	...	balance.

Treatment

As for silicon bronze (above) but finally plunge one DEOXIDISING TUBE E3 per 50 kg melt.

If die casting use ALBRAL 3.

Recommended Casting Temperatures

Light castings, under 13 mm section	1250°C
Medium castings, 13-38 mm section	1200°C
Heavy castings, over 38 mm section	1150°C

For the most effective gas removal plunge LOGAS 50 briquettes.

COPPER NICKEL ALLOYS

For example, **Nickel bronze**—20% Ni, balance copper.
Cupro nickel—66% Ni, balance copper.

Treatment

Melt down under an oxidising cover of CUPREX 1 blocks. For pressure tight castings, plunge REGENEX or LOGAS 50 briquettes when approaching the pouring stage. Before pouring plunge DEOXIDISING TUBES NS together with DEOXIDISING TUBES MG(R). Finally, if degassing by means of REGENEX or LOGAS 50 has not been carried out, plunge DEOXIDISING TUBES L, containing lithium.

Quantity Required per 50 kg Melt

1 kg CUPREX 1 blocks.

250 g REGENEX.

One DEOXIDISING TUBE NS4 + 1 DEOXIDISING TUBE MG(R) 10.

Two DEOXIDISING TUBES LO (if degassing not carried out previously.)

Recommended Pouring Temperatures

	Nickel bronze	Cupro nickel
Light castings under 13 mm section	1400°C	1560°C
	2550°F	2840°F
Medium castings, 13-38 mm section	1350°C	1530°C
	2460°F	2790°F
Heavy castings over 38 mm section	1280°C	1500°C
	2335°F	2730°F

NICKEL-SILVER ALLOYS

(Nickel Brass)

For example, 18% Ni, 10% Zn, 2½% Sn, up to 7% Pb, balance copper.

Treatment

Melt down under a flux cover of CUPREX 1 blocks. Before pouring plunge DEOXIDISING TUBES E followed by DEOXIDISING TUBES DS. The former provides the manganese deoxidant whilst the latter introduces phosphorus.

Quantity Required per 50 kg Melt

0·5 kg CUPREX 1 blocks.
Two DEOXIDISING TUBES E3 ⎫ or one DEOXIDISING TUBE NS4.
One DEOXIDISING TUBE DS4 ⎭ See notes at end of this section.

Recommended Pouring Temperatures

Light Castings, under 13 mm section	1300°C
Medium Castings, 13-38 mm section	1280°C
Heavy Castings, over 38 mm section	1250°C

Nickel Additions

It is often necessary to introduce further quantities of nickel into melts, either nickel bronze or nickel-silver alloy. Nickel is best added in the form of 50/50 copper-nickel shot such as DENSITE.

Alternative Deoxidants

Considering the chemical reactions involved in the deoxidation of cupro-nickel and nickel-silver alloys, the ideal deoxidant should function as follows :

1. It should combine with all oxygen present and reduce or combine with any oxides to form a fluid slag.

2. Deoxidation products should not be entrained in the solidified casting.

3. A residual deoxidant should not adversely affect physical or working properties of the metal and should prevent further oxidation during pouring.

Phosphorus seems to satisfy all of the above requirements but a residual content of 0·025% is regarded necessary to ensure adequate deoxidation. With phosphorus at or above this level, difficulties may arise during the working of the nickel silver or high nickel bearing alloys, due to embrittlement.

Alternative and readily available deoxidising agents are as follows :

MAGNESIUM

This is very effective, particularly as it eliminates the harmful effects of sulphur, but the oxide formed tends to remain entrapped in the metal at grain boundaries, thereby causing embrittlement.

MANGANESE

This is an excellent deoxidant and it also reacts with sulphur to some extent. It is present in DEOXIDISING TUBES E. Manganese is known to impart some grain refinement and it is also believed to effect some improvement in metal behaviour during rolling.

CALCIUM

This is considered a very good deoxidant, although metal fluidity is slightly reduced.

SILICON

This element could be classed as a satisfactory deoxidant, although it forms a solid oxide. Separation of the oxide from the metal is, however, rapid.

BORON

This should also be considered as a satisfactory deoxidant, particularly in view of its possibilities as a grain refining agent.

DEOXIDISING TUBES NS contain balanced proportions of many of the deoxidants mentioned above. They are a recent FOSECO development and are particularly recommended for producers of nickel-silver castings which are to be subsequently worked by rolling, etc. Tests with DEOXIDISING TUBES NS have brought about a significant reduction in rejects in the wrought nickel-silver industry. In the melting recommendations previously given for nickel-silver alloys, DEOXIDISING TUBES NS should therefore be used instead of the standard recommendations given, which are applicable to sand castings production.

For the deoxidation of cupro-nickel alloys, magnesium stick is widely used in view of its deoxidation efficiency. It is suggested, nevertheless, that DEOXIDISING TUBES NS are used together with magnesium metal on an approximately 50/50 basis as given in previous recommendations.

SECTION IV

IRON CASTINGS

COMPOSITION OF TYPICAL CUPOLA CHARGE MATERIALS

Pig Iron. Grading of pig iron by fracture has now largely given place to grading by chemical analysis. The appearance of the fracture of sand cast pig iron is governed chiefly by the silicon content. Although the modern foundryman buys his pig irons to chemical specification, the old grading numbers still persist.

Silicon contents are fairly consistent throughout the grade numbers of different makers but the elements manganese and phosphorus vary to a considerable degree. The analyses given below are representative of the usual grade numbers with respect to silicon, which decreases as the grade number rises. Throughout any one make of pig iron the manganese and phosphorus do not change much.

Grade No.	Total Carbon percentage	Graphite percentage	Combined Carbon percentage	Silicon percentage	Manganese percentage	Sulphur percentage	Phosphorus percentage
Silky ...	3·0 to 3·3	3·0 to 3·26	0·04	4·0 to 5·0	1·0 to 1·4	0·02	1·0 to 1·5
No. 1 ...	3·8	3·7	0·1	3·0 to 4·0	1·0 to 1·4	0·03	1·0 to 1·5
No. 2 ...	3·7	3·5	0·2	2·5 to 3·5	1·0 to 1·4	0·035	1·0 to 1·5
No. 3 ...	3·65	3·3	0·35	2·75′	1·0 to 1·4	0·04	1·0 to 1·5
No. 4 ...	3·5	3·1	0·4	2·0 to 2·5	0·9 to 1·3	0·05	1·0 to 1·5
No. 4 forge ...	3·3	2·8	0·5	2·25	0·8 to 1·2	0·07	1·0 to 1·5
No. 5 ...	3·25	2·6	0·65	1·75	0·7 to 1·0	0·1	1·0 to 1·5
Mottled ...	3·1	1·6	1·5	1·0	0·5	0·12	1·0
White ...	3·25	0·05	3·2	0·6	0·3	0·15	0·8

Hematite pig irons are graded similarly, but their special feature is very low sulphur and phosphorus, usually under 0·05% for each element.

Cylinder and refined pig irons are graded according to their total carbon and silicon contents which are adjusted to give maximum strength and density of structure with good machining properties.

Scrap. Most cupola charges contain pig iron and scrap in various proportions. Scrap cast iron is graded under a number of headings. Although its composition may vary over fairly wide limits some average compositions are as follows :

Grade	Total Carbon percentage	Silicon percentage	Manganese percentage	Sulphur percentage	Phosphorus percentage
Machinery	3·5	2·0	0·55	0·1	0·8
Railroad	3·3	2·0	0·60	0·1	0·7
Agricultural	3·5	2·25	0·60	0·09	1·0
Light	3·6	2·3	0·50	0·09	1·2
Malleable	2·0	0·75	0·30	0·15	0·15
Mild Steel	0·25	0·20	0·40	0·05	0·05

Sizes and Weights of Scrap for Cupolas. Charge materials should be within definite size limits which vary according to the internal diameter of the cupola. Uniform melting and working of the furnace depends greatly on correct charge weights and suitably graded components. Speaking generally, it is safest, if "hanging" up of charges is to be avoided, to restrict the overall length of the pig iron and scrap to a dimension equal to one-third of the internal diameter of the cupola. The appended tables give maximum dimensions which should not be exceeded.

Sizes and Weights of Scrap for Cupolas

Cast Iron

Diameter Cupola inside lining	Maximum Thickness (mm)	Minimum Thickness (mm)	Maximum Length (mm)	Maximum Weight (kg)
30″	100	6	450	50
48″	100	6	600	50
60″	150	6	900	75
72″	150	6	1200	75
90″	200	6	1500	100

Diameter Cupola inside lining	Maximum Thickness (mm)	Minimum Thickness (mm)	Maximum Length (mm)	Maximum Weight (kg)
30″	25	6	450	13
48″	25	6	600	22
60″	50	6	900	34
72″	75	6	1200	45
90″	100	6	1500	67

Coke. Cupola coke should be dense, hard, strong and of a silvery grey colour. It should be within the limits of composition shown below.

Fixed carbon not less than 90%
Ash not more than 10%
Sulphur not more than ... ... 1·0%

Some compositions typical of different kinds of coke are as follows :

	By-Product	Beehive	Patch Petroleum
Fixed carbon ...	90·0 to 92·0	93·0 to 95·0	98·0 plus
Ash	7·0 to 9·0	5·0	0·50
Sulphur	0·6	0·5	0·30
Volatile matter ...	0·6 to 0·9	0·9	1·0

Limestone. A good grade of Limestone suitable for fluxing the cupola should contain :

96 to 98% calcium and magnesium carbonates.
Less than 2% silica.

Cupola Melting Gains and Losses. In calculating metal charges the following average allowances should be made :

Silicon 10 to 15% loss.
Manganese 20 to 30% loss.
Sulphur 20 to 50% gain.
Phosphorus 1 to 2% gain.

These changes depend upon the conditions of operating the cupola and therefore may vary widely. Under properly controlled working the gains and losses occur within fairly well defined limits as shown.

The silicon loss applies to silicon contents between 1·5% to 3·5%. When silicon falls below 1% the melting loss is greater and may be as much as 50% of the silicon charged.

INFLUENCE OF THE ELEMENTS NORMALLY PRESENT IN CAST IRON ON THE PHYSICAL PROPERTIES

Element	Fluidity	Softness	Shrinkage	Strength	Density	Chill	Sulphur	Combined Carbon	Graphite
Combined Carbon	Decreases	Decreases	Increases	Increases	Increases	Increases	Neutral	–	Decreases
Graphite	Increases	Increases	Decreases	Decreases	Decreases	Decreases	Neutral	Decreases	–
Silicon	Increases	Increases	Decreases	Decreases	Decreases	Decreases	Decreases	Decreases	Increases
Manganese	Increases	up to 1% Increases	Little effect	Increases	Increases	above 1% Increases	Decreases	Increases	Decreases
Sulphur	Decreases	Decreases	Increases	Decreases	Increases	Promotes	–	Increases	Decreases
Phosphorus	Increases	Decreases	Aggravates	Decreases	Neutral	Little effect	Neutral	Tends to Increase	Neutral

INFLUENCE OF NORMAL CONSTITUENTS IN CAST IRON

When deciding upon a chemical composition for any given class of casting the effects of the main constituents have to be considered. These, summarised briefly, are as follows :

Carbon. In all cast iron compositions carbon is the most important element. Its total amount and state of existence in the structure, that is combined as carbide or free as graphite, exert a major influence on the physical properties.

For alloys of iron and carbon only, the eutectic, or lowest freezing point composition is that having 4·3% of carbon. With carbon contents above this figure the alloys are termed hyper-eutectic, and those below it hypo-eutectic alloys. The hyper-eutectic compositions are liable to have coarse open-grained structures and comparatively large flakes of "kish" graphite when cooled slowly as in ordinary sand moulds. On the other hand the hypo-eutectic alloys have relatively fine graphite and denser, tighter structures. The tendencies mentioned become more extreme as the carbon ratio rises in the hyper-eutectic, or falls in the hypo-eutectic compositions.

Total carbon content also has an important influence on liquid shrinkage and solid contraction. Irons containing less than 3% become more difficult to cast sound as the carbon falls below this value and methods of running and feeding approaching those used in steel casting practice become more necessary.

Commercial cast irons are not simply alloys of iron and carbon, but invariably contain proportions of silicon, manganese, phosphorus and sulphur as well. The presence of these elements affects the carbon solubility. It is known that silicon and phosphorus each reduce the amount of carbon required to form the eutectic alloy by 0·3% for each 1% of silicon and 1% of phosphorus present.

Therefore, to find the carbon equivalent value for any composition add together the silicon and phosphorus percentages and divide by 3. The result so found is added to the carbon percentage. If this then shows greater than 4·3, the composition will tend to give open-grained castings with large graphite flakes, but if less than 4·3 the structure will be denser, sounder and contain less graphite as the percentage falls from the eutectic value.

For example, assuming a composition having total carbon 3·1%, silicon 2·0% and phosphorus 1·0%. The carbon equivalent found as explained is 2·0 + 1·0 = 3·0 ÷ 3·0 = 1·0. Then 3·1 + 1·0 = 4·1, which is below the 4·3 value. Irons in the high duty class have carbon equivalent values of about 3·5 to 3·8.

Silicon. Next in importance to carbon, with respect to influence on the properties of cast iron, is silicon. The presence of silicon tends to throw the carbon out of combination or solution in the iron and make it form graphite. In so doing, it reduces both the hardness and the shrinkage and decreases strength and density. For the best mechanical properties with good resistance to pressure and wear carbon and silicon ratios must be decided in relation to each other. That is the silicon must be decreased as the total carbon rises and vice versa, so as to maintain carbon equivalent values within the desired limits.

Manganese. First neutralises the ill effect of sulphur by combining with it to form manganese sulphide in place of iron sulphide. Manganese sulphide is lighter than iron sulphide and so tends to float out of the molten iron as slag. Iron sulphide is retained in the melt and deposited around the grain boundaries rendering the metal hot short and likely to produce cracked castings.

Manganese in excess of that required to combine with the sulphur forms manganese carbide and increases the hardness and chilling tendency. As sulphur is a more powerful hardener than manganese, the first effect of adding manganese often causes softening due to removal of sulphur hardness. Also in neutralising sulphur, manganese raises the fluidity of molten cast iron.

In order to ensure a correct manganese-sulphur balance in grey irons, the percentage of manganese should be at least equal to the sulphur content x 1·7 plus 0·3%. Thus, with 0·1% sulphur, the minimum manganese content is 0·1 x 1·7 = 0·17 plus 0·3 = 0·47%. Manganese up to 1·0% improves the density and strength of grey cast irons.

Sulphur. This element exerts the opposite effect to silicon, tending to prevent carbon deposition as graphite. Therefore, it has a marked hardening influence.

Apart from this, as explained under "Manganese," unbalanced sulphur makes molten iron sluggish, encouraging short running and pinhole defects due to entrapment of mould and other gases.

Phosphorus. The information given here is drawn largely from the summarised work of the British Cast Iron Research Association and refers to plain cast irons not specially alloyed, heat treated or ladle treated.

Mechanical strength is at a maximum at about 0·35% phosphorus. At about 0·65% the strength properties fall and continue to do so as the phosphorus further increases.

The Brinell hardness increases about four points for each 0·1% phosphorus added to a non-phosphoric iron.

Impact or shock strength diminishes with increase in phosphorus content.

Speaking generally, phosphorus increases wear resistance, other things being equal.

It is well known that phosphorus increases the fluidity and running power of all cast iron compositions. At the same time it renders more difficult the production of sound castings where ribs, internal bosses and abrupt variations in sectional thickness are encountered.

Compositional limits for adequate strength and soundness in phosphoric cast irons taking into account both silicon and total carbon, have been worked out by the British Cast Iron Research Association.

CUPOLA OPERATION DATA (METRIC)

Inside Diameter at Melting Zone (cm)	Area at Melting Zone (cm²)	Total Tuyere Area (cm²)	Melting Rate per hour tonnes	Coke Charge (kg)	Iron Charge (kg)	Limestone Charge (kg)	Blast Volume (m³ per min)	Blast Pressure (cm W.G.)	Bed height above Tuyeres (cm)	Fan force de cheval	Height from Tuyeres to Charge Door Sill (metres)
46	1640	320	1·0	11	100	3·6	17·5	30	100	2·0	2·5
61	2900	645	2·0	20	200	6·3	31	30	100	5·0	3·0
76	4500	1130	3·0	32	320	10·4	48	35	105	7·0	3·0
91	6550	1290	4·5	45	450	15	69	38	105	12·0	3·5
107	8900	1800	6·0	60	600	20	93	40	110	17·0	3·5
122	11 700	1930	8·0	82	820	27	123	43	110	23·0	4·0
137	14 700	2450	10·0	101	1010	32	156	45	110	30·0	4·0
152	18 300	2570	12·5	125	1250	41	191	48	110	40·0	4·5
168	22 000	3150	15·0	158	1580	50	232	51	110	50·0	5·0
183	24 200	3225	18·0	180	1800	60	278	51	110	60·0	5·0
198	31 000	3850	21·0	210	2100	68	326	51	110	70·0	5·5
213	35 800	4450	25·0	250	2500	82	382	53	115	85·0	6·0
229	41 000	5100	28·5	280	2800	90	439	56	115	110·0	6·0

The above figures represent good average practice. They are by no means critical values and can vary considerably according to individual conditions and circumstances. Their main purpose here is to act as a guide in cases where such help is required.

CUPOLA OPERATION DATA (AVOIRDUPOIS)

Inside Diam. at Melting Zone inches	Area at Melting Zone sq. in.	Total Tuyere Area sq. in.	Melting Rate per Hour tons	Coke Charge lbs.	Iron Charge	Lime-stone Charge lbs.	Blast Volume cu. ft. per min.	Blast Pressure inches W.G.	Bed Height above Tuyeres inches	Fan H.P.	Height from Tuyeres to Charge door sill
18	254	50	1·0	25	250	8	620	12	38	2·0	8' 0"
24	452	100	2·0	45	450	14	1100	12	39	5·0	9' 0"
30	707	175	3·0	70	700	23	1700	14	41	7·0	10' 0"
36	1018	200	4·5	100	1000	33	2450	15	41	12·0	11' 0"
42	1385	280	6·0	130	cwts. 12	44	3300	16	42	17·0	12' 0"
48	1810	300	8·0	180	16	60	4350	17	42	23·0	13' 0"
54	2290	380	10·0	225	20	70	5500	18	43	30·0	14' 0"
60	2827	400	12·5	275	24	90	6750	19	43	40·0	15' 0"
66	3421	490	15·0	350	30	110	8200	20	44	50·0	16' 0"
72	4072	500	18·0	400	35	130	9800	20	44	60·0	17' 0"
78	4778	600	21·0	465	40	150	11500	20	44	70·0	18' 0"
84	5542	690	25·0	540	48	180	13500	21	45	85·0	19' 0"
90	6362	790	28·5	620	55	200	15500	22	45	110·0	20' 0"

The above figures represent good average practice. They are by no means critical values and can vary considerably according to individual conditions and circumstances. Their main purpose here is to act as a guide in cases where such help is required.

INFLUENCE OF SOME ALLOYING ELEMENTS USED IN THE PRODUCTION OF CAST IRON UPON ITS STRUCTURE

Element	Percentage most frequently used	Effect on Chill	Effect on Structure	Comments
Aluminium	Up to 2·0	Reduces. 1·0% is approximately equivalent to 0·5% Silicon.	Stabilises Ferrite. Increases and coarsens the Graphite. Decreases the hardness.	Generally used in small percentages as a de-oxidant and scavenger only.
Chromium	0·15 to 1·00	Increases. 1·0% approximately neutralises the graphitising effect of 1·0% Silicon.	Stabilises Cementite. Reduces and refines the Graphite. Increases the hardness.	Used for hardness, chilling power and wear resistance.
Copper	0·5 to 2·0	Decreases. 1·0% is approximately equivalent to 0·35% Silicon. Assists in control of chill depth.	Tends to increase and refine the Graphite.	Toughens the matrix and increases the fluidity.
Manganese	0·3 to 1·25	By first combining with Sulphur it tends to reduce the chill. In excess of this amount it increases the chill. 1·0% of Manganese neutralises about 0·25% Silicon.	Stabilises Austenite. Refines the Graphite and Pearlite.	Also acts as a deoxidiser. Gives grain refinement, density and increased fluidity.
Molybdenum	0·30 to 1·00	Increases. 1·0% is as effective as about 0·33% Chromium and neutralises the effect of 0·35% Silicon.	Refines the Graphite and Pearlite.	Used chiefly in combination with Nickel. Copper and Chromium in the production of high strength irons.

Influence of Some Alloying Elements used in the Production of Cast Iron upon its Structure—*continued*

Element	Percentage most frequently used	Effect on Chill	Effect on Structure	Comments
Nickel	0·10 to 3·00	Decreases, 1·0% is about equal to 0·33% Silicon and offsets the chilling effect of about 0·33% Chromium.	Stabilises Austenite. Refines the Pearlite and Graphite.	Improves the density and toughness. Evens out the hardness between light and heavy sections.
Silicon	0·5 to 3·50	Reduces.	Stabilises Ferrite. Increases the quantity and coarseness of the Graphite.	Softens, weakens and imparts an open grained structure.
Titanium	0·05 to 0·10	Decreases powerfully.	Increases but refines the Graphite.	Used chiefly as a deoxidiser and degasser. Improves fluidity.
Vanadium	0·15 to 0·50	Increases strongly. 1·0% Vanadium offsets the chill reducing influence of about 1·75% Silicon.	Stabilises Cementite and improves the structure of the chill.	Increases hardness and resistance to wear and heat.
Zirconium	0·10 to 0·30	Mildly reduces.	Assists formation of Graphite.	Reduces hardness. Deoxidises and improves the fluidity and density.

CUPOLA CHARGE CALCULATIONS

A wide variety of materials can be used in order to obtain almost any required analysis. In consequence it is largely a matter of experience to determine whether the total carbon content will show a gain or a loss on the calculated content and to what degree.

Under the normal conditions of melting the tendency is for the total carbon to attain the eutectic equivalent. If the quantity charged is above this value, a loss may be expected. On the other hand, where the charge contains less than the eutectic value, the trend is towards a carbon pick-up. For total carbon contents between 3·2% and 3·5% as charged, little change is likely to occur during melting. In view of the element of judgment required in assessing total carbon it will be omitted from the calculations given as examples.

Suppose the materials available are a No. I grade foundry pig iron and some cast iron scrap of the agricultural type. Chemical analyses of these two materials are as follows :

			Silicon	Manganese	Sulphur	Phosphorus
Pig iron	...	...	3·5	1·2	0·03	1·0
Scrap	...	...	2·25	0·6	0·09	1·0

It is required to make up a charge from these two materials which on melting will have a composition between the following specified limits :

Silicon	...	...	2·20 to 2·4%
Manganese	...	...	0·60 to 0·80%
Sulphur	...	...	0·08 to 0·10%
Phosphorus	...	...	0·75 to 1·00%

Taking silicon as the controlling element, it is seen that the two items charged in equal proportions yield a silicon content of 2·87% or allowing a 10% melting loss, say 2·6%, which is above the top limit specified.

Rough preliminary trials along these lines will soon suggest likely proportions before going to the trouble of making the calculation in greater detail. Thus as a charge composed of equal proportions of pig iron and scrap iron gives too high a silicon content it is obvious that less than 50% of the pig iron and more than 50% of the scrap iron must be taken. As the No. I pig iron has a good scrap carrying capacity by reason of its high silicon percentage, advantage can be taken of this to try out a mixture of 30% pig iron and 70% scrap. The full calculation is then as follows :

Ratio	Material	Silicon	Silicon × per cent	Manganese	Manganese × per cent	Sulphur	Sulphur × per cent	Phosphorus	Phosphorus × per cent
30%	No. 1 Pig Iron	3·5	105·0	1·2	36·0	0·03	0·9	1·0	30·0
70%	Cast Iron Scrap	2·25	157·5	0·6	42·0	0·09	6·3	1·0	70·0
100%	Totals		262·5		78·0		7·2		100·0
	Divide by 100		2·625		0·78		0·072		1·0
	Adjust for gains and losses		−0·26		−0·2		+ 0·022		Little change
	Theoretical Analysis		2·36		0·58		0·094		1·0

The theoretical chemical analysis thus comes to be :

Silicon ... ... 2·36% allowing for a 10% loss.
Manganese ... ... 0·58% ,, ,, 25% ,,
Sulphur ... ... 0·094% ,, ,, 30% gain.
Phosphorus ... ... 1·00% ,, ,, no change.

These results are within the specified requirements, except perhaps for the manganese, which is a little low. This could be brought up by adding manganese to the charge in the form of ferro-alloy briquettes, or by means of manganese ladle additions. Silicon and phosphorus can, where necessary, be increased in a similar manner.

CURTAILMENT OF SULPHUR PICK UP DURING CUPOLA MELTING OF IRON CHARGES

High Scrap Ratios

When conditions make it necessary to use high proportions of returned or purchased scrap in the cupola charges, the sulphur content of the resulting cast iron tends to become high. It is therefore desirable that all steps should be taken to keep the sulphur pick up within safe and reasonable limits. In grey cast irons sulphur fulfils no useful purpose but its presence is unavoidable and it exerts little ill effect provided it is not allowed to increase unduly and is offset by an adequate percentage of manganese or other element capable of neutralising its adverse effect.

Detrimental Effect of High Sulphur

Sulphur in cast iron exerts a carbide stabilising or chill inducing tendency. Therefore, if it is allowed to increase without check it will produce irons which are sluggish when molten and give rise to trapped gases and unmachinability in the castings. Irons with a high unbalanced sulphur are also liable to crack in thin sections, and may show inverse chill. They are most unsuitable for castings which are required to be subsequently vitreous enamelled because they tend to produce blisters and pin-holing in the enamel coating.

Conditions Favouring High Sulphur Pick Up

Every time pig iron or scrap iron is melted in the cupola the sulphur content increases. Most of the additional sulphur is picked up from the coke and this is especially the case if the coke is of poor quality and high in sulphur.

Oxidising conditions of melting are favourable to a high sulphur pick up. This is because ferrous oxide is formed which passes into the slag, and a slag of high iron content will not take up sulphur.

Oxidising Conditions of Melting

These result when the iron charge melts too near the mouth of the tuyeres. A number of causes are responsible for this, namely :

1. Bed coke too low, either as charged originally, or allowed to become so during the course of the blow.

2. Use of too heavy a metal charge for the size of the cupola.

3. Soft coke of a highly reactive nature and coke of small size which allows the bed to burn away too rapidly.

4. Too little coke used in between each metal charge.

It is also possible to obtain oxidised metal by using in the charges a heavy proportion of badly rusted or burnt scrap iron.

Curtailment of Sulphur Pick Up by Correct Fluxing Technique

The chemistry of fluxing reactions is highly complex, and much has yet to be learned of the way in which slag forms. Neither lime nor magnesia, by themselves, melt at temperatures generated within the cupola. It is chiefly as a result of their combination chemically with silica that they form liquid slags. Such reactions do not take place until the materials reach the melting zone.

Therefore, it is of importance to make additions to the flux such as will allow fluxing reaction to start somewhat higher up in the cupola stack. By so doing, liquid flux will trickle down over the coke and limestone, thereby speeding up combustion and refining throughout the melting zone. At the same time, a hotter and more active slag will be produced. In this way, the coke is constantly flushed in the melting zone, allowing more intimate contact of the blast and of the droplets of iron with the surface of the incandescent fuel. Thus, better melting is promoted and higher temperatures generated which result in an increased carbon pick up and molten iron of high fluidity and freedom from contained gas.

The Function of BRIX 555 blocks

Although individual cases may need special consideration, normally, a limestone addition of about 3% is common practice but this can be reduced by one-third approximately, i.e., to 2% when introducing BRIX to the cupola. For the first half dozen charges BRIX should be added at the rate of 1·5 kg per 1000 kg. Thereafter a lower regular input will suffice and about 1 kg per 1000 kg is suggested as a general guide. This will ensure a clean drop and produce better quality iron during the blow down.

BRIX blocks activate the fluxing reactions and speed up the rate of combustion by fluxing away coke ash and leaving a cleaner coke surface at the face of which combustion takes place. So the melting and superheating zone is restricted to a shorter, hotter area giving a higher temperature and reducing the time during which the descending droplets of iron are exposed to the blast in their passage to the well of the cupola. Melting losses are also reduced due to the tendency for a protective coating of reactive slag to surround the molten drops as they trickle through the coke bed.

BRIX blocks are so constituted that they react with silica to release calcium oxide. The calcium oxide is formed in the nascent condition, and is therefore extremely active as a flux. The beneficial effect of BRIX in this respect is therefore two-fold, it eliminates silica directly and also provides a particularly active fluxing agent. It is easy to realise its advantage when it is necessary to use high ash coke and poor grade limestone.

Additional benefits resulting from carrying out the fluxing technique described here, are the increased fluidity of the slag, which contributes to greater ease and smoother working of the furnace. For example, it is easier to keep the tuyeres clean and free from slag curtains. At the end of the blow the drop will be easier and cleaner, and a lower content of entrapped fine particles of metallic iron will be lost in the slag.

Maintenance of a Slag Blanket

Higher sulphur contents are often obtained in the iron resulting from the first one or two charges melted. This is because the molten iron first produced absorbs sulphur from the bed coke. It is possible to minimise this sulphur pick up by charging limestone and BRIX blocks along with the topmost layers of the bed coke. For this purpose it is recommended that three times the amount of limestone and BRIX blocks normally employed between each charge should be added to the bed coke.

Without a flux in the bed coke no appreciable quantity of slag is formed until after the first few charges have melted. Therefore, the first molten iron to collect in the well of the furnace is not only higher in sulphur but it is exposed to the impingement of the blast from the tuyeres and is often delivered lower in temperature than charges melted subsequently and protected by a slag layer. By ensuring preliminary slag formation by fluxing the bed coke, not only is the first iron melted lower in sulphur but it is also higher in temperature.

If the maximum desulphurising effect of the slag resulting from the flux is desired it should be the aim to maintain a slag layer or blanket constantly in the well of the cupola. As pointed out, this protects the molten iron from the action of the blast and promotes a further refining action in that all the droplets of molten iron have to pass through it. On long blows it will be sufficient to run off the surplus slag periodically and thereby obtain the additional benefit which is lost where the slag is allowed to run away continuously.

CHEMICAL COMPOSITION OF UNALLOYED IRONS SUITABLE FOR DIFFERENT CLASSES OF CASTINGS

Where it is intended to make alloy additions to obtain improved properties, the analysis given may be taken as that of the base iron. To obtain the maximum benefit from the alloys it is intended to add, it will be necessary to adjust the base composition, before making the additions, so as to balance it with respect to the influence of the elements to be added. The information necessary for this purpose is given in the table on pages 245 and 246.

Class of Casting	Size	Total Carbon	Silicon	Man-ganese	Sulphur	Phos-phorus
Acid Resisting ...	Thin	3·25	2·0	0·75	0·05	0·4
	Medium	3·00	1·5	1·0	0·05	0·3
	Thick	3·00	1·0	1·25	0·05	0·2
Agricultural ...	Thin	3·75	3·0	0·60	0·06	1·3
	Medium	3·50	2·5	0·70	0·08	1·0
	Thick	3·25	2·0	0·80	0·10	0·75
Air Cylinders ...	Thin	3·40	1·8	0·75	0·08	0·6
	Medium	3·20	1·50	1·0	0·1	0·4
	Thick	3·0	1·0	1·0	0·1	0·3
Annealing Boxes ...	1″ sect.	2·9	0·75	0·5	0·08	0·2
Balls for Mills ...	Medium	3·0	0·90	1·0	0·15	0·6
Bed Plates	Thin	3·5	2·0	0·70	0·08	1·0
	Medium	3·3	1·75	0·75	0·10	0·75
	Thick	3·2	1·5	0·8	0·10	0·50
Brake Shoes ...	Medium	3·25	1·5	1·50	0·10	0·60
Car Wheels ...	Thick	3·0	1·0	1·0	0·12	0·3
Do.—Chilled ...	Thick	3·3	0·7	0·8	0·13	0·3
Caustic Pots ...	Medium	3·25	1·25	0·75	0·08	0·3
Chills	Medium	3·5	2·5	0·75	0·05	0·2
Chilled Rolls ...	Small	3·3	0·8	0·3	0·05	0·5
	Medium	3·2	0·7	0·3	0·06	0·4
	Large	3·0	0·6	0·3	0·07	0·3
Couplings	Medium	3·3	1·75	0·7	0·08	0·5
Crusher Jaws ...	Medium	3·4	1·0	1·0	0·15	0·5
	Large	3·2	0·8	1·2	0·15	0·4

Chemical Composition of Unalloyed Iron Suitable for Different Classes of Castings—*continued*

Class of Casting	Size	Total Carbon	Silicon	Man-ganese	Sulphur	Phos-phorus
Cylinders—Ammonia	Medium	3·0	1·5	1·0	0·1	0·8
Do.—Automobile	Small	3·25	1·9	0·8	0·1	0·4
	Medium	3·0	1·75	1·0	0·1	0·3
Do.—Gas Engine and Diesel ...	Small	3·5	2·0	0·6	0·08	0·5
	Medium	3·25	1·50	0·8	0·09	0·4
	Large	3·0	1·25	1·0	0·10	0·3
Cylinders—Hydraulic	Medium	3·25	1·5	0·8	0·1	0·40
	Thick	3·0	1·0	1·0	0·12	0·30
Do.—Locomotive ...	Medium	3·5	1·75	0·8	0·09	0·75
Do.—Steam ...	Thick	3·25	1·5	1·0	0·1	0·50
Do.—Motor Cycle	Thin	3·3	1·8	1·0	0·1	0·80
Dynamo Frames ...	Small	3·25	3·0	0·5	0·05	0·60
	Large	3·00	2·5	0·5	0·06	0·50
Electric Work ...	Thin	3·40	3·25	0·5	0·06	1·25
	Medium	3·25	2·50	0·6	0·08	1·00
Engine Frames ...	Medium	3·50	2·0	1·0	0·08	1·00
	Large	3·25	1·5	1·0	0·1	0·75
Fan Cases	Medium	3·3	2·25	1·0	0·08	1·0
Fly Wheels ...	Thin	3·4	2·0	0·6	0·07	0·7
	Medium	3·2	1·75	0·8	0·08	0·5
	Thick	3·0	1·25	1·0	0·10	0·3
Friction Clutches ...	Thin	3·6	2·75	0·6	0·06	0·7
	Medium	3·5	2·50	0·7	0·08	0·5
Gears	Small	3·5	2·25	1·0	0·08	1·0
	Medium	3·25	1·75	1·0	0·09	0·7
	Large	3·00	1·50	1·0	0·10	0·5
Glass Moulds ...	Small	3·50	2·50	0·5	0·07	0·5
	Medium	3·30	2·0	0·75	0·08	0·3
	Large	3·20	1·75	1·0	0·10	0·2
Grate Bars	1″ Thick	3·20	1·25	1·0	0·12	0·3
Grinding Balls ...	Small	3·25	1·5	0·3	0·10	0·3
Grinding Plates ...	Medium	3·5	0·6	0·6	0·12	0·3
Hardening Pots ...	Thin	3·2	1·0	0·6	0·06	0·2
Hardware ...	Thin	3·5	2·75	0·6	0·08	1·5
Heat Resisting ...	Medium	3·3	1·6	1·0	0·08	0·25
	Thick	3·0	1·2	1·0	0·10	0·10

Class of Casting	Size	Total Carbon	Silicon	Manganese	Sulphur	Phosphorus
Hollow Ware ...	Thin	3·5	2·7	1·0	0·07	1·5
Ingot Moulds ...	Medium	3·6	1·5	1·0	0·06	0·1
	Large	3·3	1·2	1·0	0·05	0·1
Machine Tools ...	Thin	3·50	2·3	0·75	0·08	1·0
	Medium	3·25	2·0	1·0	0·1	0·75
	Thick	3·00	1·2	1·0	0·1	0·50
Mine Car Wheels ...	Medium	3·1	1·0	0·75	0·1	0·40
Mowers	Thin	3·5	2·5	0·50	0·1	1·0
Ornamental Work	Thin	3·6	3·0	0·50	0·05	1·25
	Medium	3·5	2·75	0·60	0·05	1·00
	Thick	3·25	2·50	0·70	0·06	0·75
Permanent Moulds	Medium	3·5	2·3	0·90	0·07	0·4
Do.—M. Castings ...	Medium	3·2	3·0	1·0	0·06	0·8
Piano Frames ...	Medium	3·4	2·5	0·8	0·08	0·6
Pipes (Water) ...	Thin	3·5	2·5	0·6	0·1	1·0
	Medium	3·3	2·25	0·8	0·1	1·0
Do.—(Steam) ...	Medium	3·25	1·50	0·9	0·07	0·3
Pistons (Automobile)	Thin	3·3	1·8	0·7	0·1	0·7
Do.—Rings... ...	Thin	3·5	2·0	0·7	0·1	0·7
	Medium	3·25	1·75	0·8	0·1	0·5
Plough Points ...	Thin	3·25	1·0	0·9	0·08	0·3
Pulleys	Thin	3·6	2·75	1·0	0·07	1·0
	Medium	3·3	2·25	1·0	0·08	0·75
	Thick	3·0	1·75	1·0	0·10	0·6
Radiators	Thin	3·5	2·5	0·7	0·06	0·8
Rolls	Medium	3·2	0·7	0·3	0·1	0·4
	Large	3·0	0·6	0·5	0·1	0·2
Soft Castings ...	Thin	3·7	2·75	0·5	0·05	0·6
	Medium	3·5	2·50	0·6	0·06	0·5
Slag Pots	Medium	3·3	1·70	0·8	0·07	0·2
Stove Plates ...	Thin	3·5	2·7	0·5	0·06	1·0
	Medium	3·25	2·25	0·6	0·08	1·0
Typewriter Frames	Thin	3·5	2·30	0·5	0·07	1·0
Valves	Thin	3·3	2·2	0·6	0·07	0·6
	Medium	3·0	1·8	0·8	0·08	0·5
	Thick	2·8	1·2	1·0	0·1	0·3
White Iron Castings	Medium	3·0	0·8	0·4	0·15	0·5

BRITISH STANDARD SPECIFICATIONS
FOR CAST IRON

Note: The information given in this section is correct at the time of going to press. However the whole of the BS specifications for all cast irons are under revision to include metrication, and not all have been reissued in time for inclusion in these notes. Part of the information is therefore in metric units and part in avoirdupois. Regrets are expressed for any inconvenience occasioned.

Chemical Composition and Mechanical Tests

None of the British Standard Specifications No. 1452/1961 covering general grey iron and high duty iron castings respectively, ask for definite chemical composition. Indeed they clearly state that "The composition of the irons as cast shall be left to the discretion of the manufacturers but the maximum percentage of phosphorus may be specified by the engineer or purchaser if he so desires." This is an advantage in that it leaves the foundry man free to use whatever materials suit him best, provided he can obtain the required mechanical test results from the cast bars. On the other hand, D.T.D. and some other specifications covering cast iron for Air Ministry requirements do stipulate limits for chemical composition as well as the mechanical tests.

British Standard 1452:1961

Grade numbers are arranged so as to indicate the tensile strength in tons per square inch required on the 1·2 in. diameter bar.

Tensile Test Bars

Cast-on test bars may be specified where the design of the casting, the grade of iron and the method of running permit.

Separately cast bars shall be poured at the same time and from the same ladle of metal as the casting or castings they represent.

Cross-sectional thickness of casting		Nominal diameter of test bar as cast	Gauge dia.	Tensile strength, minimum						
				Grade						
Over	Up to and including			10 (See Note)	12	14	17	20	23	26
in.	in.	in.	in.	tons/in.^{2}	tons/in.^{2}	tons/in.^{2}	tons/in.^{2}	tons/in.^{2}	tons/in.^{2}	tons/in.^{2}
—	3/8	0·6	0·399	11·0	13·0	16·0	19·0	22·0	25·0	28·0
3/8	3/4	0·875	0·564	10·5	12·5	15·0	18·0	21·0	24·0	27·0
3/4	1 1/4	1·2	0·798	10·0	12·0	14·0	17·0	20·0	23·0	26·0
1 1/4	1 5/8	1·6	1·128	9·5	11·5	13·5	16·0	19·0	22·0	25·0
1 5/8	—	2·1	1·493	9·0	11·0	13·0	15·0	18·0	21·0	24·0

NOTE. Although mechanical properties are specified for Grade 10, the material will be tested for these properties only when specially required by the purchaser.

Standard Transverse Test Bars

Transverse test bars have standard diameters adjusted to relate to the average sectional thickness of the casting represented, as indicated in table below.

Dimensions of Transverse Test Bars related to main cross-sectional thickness of casting

Diameter of Test Bar (in.)	Overall Length (in.)	Main cross-sectional thickness of casting represented (in.)
0·6	10	Not exceeding $\frac{3}{8}$
0·875	15	Over $\frac{3}{8}$ and not exceeding $\frac{3}{4}$
1·2	21	Over $\frac{3}{4}$ and not exceeding $1\frac{1}{8}$
1·6	21	Over $1\frac{1}{8}$ and not exceeding $1\frac{5}{8}$
2·1	27	Over $1\frac{5}{8}$

For castings of more than 2 inches main sectional thickness a test bar of larger diameter than 2·1 inches may be used by agreement between the manufacturer and purchaser.

Formerly tensile and transverse tests were on an equal basis. This no longer applies, the transverse test being subsidiary.

Transverse Test

Transverse test requirements are specified as breaking load in pounds and as transverse rupture stress. The latter is used rather than the actual breaking load, because it takes into account the variations in span and diameter, thus making the test results more capable of direct comparison when obtained from bars of different dimensions. Variations due to rate of cooling, however, cannot be allowed for in the calculations.

Transverse rupture stress is obtained from the formula

$$\frac{\text{Breaking load in tons x span in inches}}{4 \times \text{modulus of section}}$$

Since we are concerned with round bars only, the modulus of section of a round bar is obtained from the expression 0·0982 x diam.^{3}, where the diameter is measured in inches. As in all standard bars, span and section for each are constant ; the formula shown above can be reduced to a single factor for each. These factors are given in the table below, and it is necessary only to multiply the transverse breaking load in lbs. by the factor for the bar concerned to obtain the transverse rupture stress in tons per square inch.

Factors for converting Transverse Breaking Load in Pounds to Transverse Rupture Stress in Tons per Square Inch for Standard Round Bars

Bar Diameter (in.)	Breaking Span (in.)	Factor
0·6	9	0·0475
0·875	12	0·0204
1·2	18	0·0118
1·6	18	0·005
2·1	24	0·00295

Deflection has an important bearing on the toughness and quality of cast iron and forms part of the transverse specification. A combined summary of the tensile and transverse requirements for all the grades is given in the table appended.

Grade	Transverse					Tensile		
	Bar Diameter Inches	Span Inches	Load lbs.	Rupture Stress tons Square Inch	*Deflex Inches	Gauge Diameter	Area	Minimum Tensile tons per Square Inch
Grade 10	0·6	9	430	20·5	Not specified	0·399	0·125	11·0
	0·875	12	975	19·9		0·564	0·25	10·5
	1·2	18	1640	19·3		0·798	0·50	10·0
	1·6	18	3750	18·7		1·128	1·00	9·5
	2·1	24	6100	18·0		1·493	1·75	9·0
Grade 12	0·6	9	485	23·1	Not specified	0·399	0·125	13·0
	0·875	12	1100	22·4		0·564	0·25	12·5
	1·2	18	1850	21·8		0·798	0·50	12·0
	1·6	18	4250	21·2		1·128	1·00	11·5
	2·1	24	6950	20·5		1·493	1·75	11·0
Grade 14	0·6	9	565	26·9	Not specified	0·399	0·125	16·0
	0·875	12	1255	25·6		0·564	0·25	15·0
	1·2	18	2060	24·3		0·798	0·50	14·0
	1·6	18	4750	23·7		1·128	1·00	13·5
	2·1	24	7830	23·1		1·493	1·75	13·0
Grade 17	0·6	9	645	30·7	0·09	0·399	0·125	19·0
	0·875	12	1440	29·4	0·13	0·564	0·25	18·0
	1·2	18	2380	28·1	0·20	0·798	0·50	17·0
	1·6	18	5390	26·9	0·16	1·128	1·00	16·0
	2·1	24	8680	25·6	0·22	1·493	1·75	15·0
Grade 20	0·6	9	725	34·4	0·10	0·399	0·125	22·0
	0·875	12	1625	33·2	0·15	0·564	0·25	21·0
	1·2	18	2700	31·9	0·23	0·798	0·50	20·0
	1·6	18	6150	30·7	0·19	1·128	1·00	19·0
	2·1	24	9970	29·4	0·25	1·493	1·75	18·0
Grade 23	0·6	9	805	38·2	0·10	0·399	0·125	25·0
	0·875	12	1815	37·0	0·15	0·564	0·25	24·0
	1·2	18	3030	35·7	0·23	0·798	0·50	23·0
	1·6	18	6900	34·4	0·19	1·128	1·00	22·0
	2·1	24	11250	33·2	0·25	1·493	1·75	21·0
Grade 26	0·6	9	885	42·0	0·10	0·399	0·125	28·0
	0·875	12	1995	40·7	0·15	0·564	0·25	27·0
	1·2	18	3350	39·5	0·23	0·798	0·50	26·0
	1·6	18	7660	38·2	0·19	1·128	1·00	25·0
	2·1	24	12540	37·0	0·25	1·493	1·75	24·0

*These values represent minimum deflections.

†For overall length see Table of Dimensions of Standard Test Bars, page **255**.

SUMMARY OF TRANSVERSE AND TENSILE TEST REQUIREMENTS
B.S. 1452/1961

RELATIONSHIP BETWEEN CARBON EQUIVALENT AND TENSILE STRENGTH OF GREY CAST IRON

BS 1452: 1956 quotes seven grades of cast iron, the basis being the minimum tensile strength of a test bar machined to standard dimensions from a 1·2 in. diameter as cast bar. There is no requirement in terms of composition of individual irons, and the foundryman is free to make his own choice based on requirements for the particular casting, section size, etc. To help in this a relationship between composition of the iron and expected tensile strength has been worked out. The elements that have the greatest effect on strength of flake irons are carbon, silicon, and phosphorus, and the influence of each is dependent on the presence of the others. The matter would be much simplified if only one factor had to be taken into consideration, and fortunately a simple equation allows this to happen. If, to the total carbon content of the iron, is added one-third of the silicon content plus one-third of the phosphorus content, a figure known as the carbon equivalent (C.E.) is obtained. This is the figure equivalent to the carbon content of a simple iron-carbon alloy if no silicon or phosphorus was present to complicate matters. The equation can be expressed as:

$$\text{Carbon Equivalent (C.E.)} = \text{T.C. \%} + \frac{\text{Si\%} + \text{P\%}}{3}$$

Since the structure (and hence the strength) of flake irons is a function of composition a knowledge of the C.E. of an iron can give an approximate indication of the strength to be expected in any sound section. This relationship is conveniently expressed in graphical form, Fig. 1:

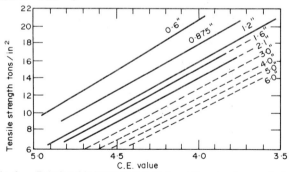

Fig. 1. Relationship between tensile strength and carbon equivalent value for various bar diameters.

Thus an iron with a C.E. value of 4·0 should give a tensile strength of 18 tons/sq. in. in a 0·875 in. diameter sand cast bar or 16 tons/sq. in. in a 1·2 in. diameter bar. It is important to remember that each line of the graph represents an average figure and a band of uncertainty exists on either side.

From Fig. 1 it is possible to construct a series of curves relating strength and section and these are shown in Fig. 2 for the five basic gradings of BS 1452:1956.

Fig. 2. Variation of tensile strength with bar diameter or section thickness for several grades of iron.

GREY CAST IRON — CLASSIFICATION OF GRAPHITE FLAKE SIZE AND SHAPE

When a polished microspecimen of a flake iron is examined under the microscope, the form, distribution, and size of the graphite can be classified. This is done by comparing the specimen at a standard magnification of 100 diameters with a stylised series of standard diagrams and allocating letters and numerals to indicate form, shape, and size of the graphite. The standard reference data

261

sheet for this purpose has been agreed by an international committee and is that proposed by the American Society for the Testing of Metals A247-47.

Certain requirements must obviously be met before a sample is evaluated. Attention must clearly be paid to the location of the microspecimen in relation to the rest of the casting, to the wall thickness, to the proximity of chills, denseners, etc., and to the distance from the as-cast surface. Some care is also necessary in grinding and polishing so that as much graphite as possible is retained in a truly representative cross-section. The specimen is normally examined in the as polished, unetched condition at a magnification of 100 diameters and with a field of view of approximately 80 mm diameter.

Graphite form is indicated by a Roman numeral, e.g. Type I.

Graphite distribution is indicated by reference to letters A-E, e.g. Type IC.

Graphite size is indicated by numbers from 1-8, e.g. Type 1C4.

An indication of the microstructure of a flake iron can therefore be given by, for example, a description such as graphite Type IIIB4, suggesting this is a rosetted graphite structure with the longest flakes between ½-1 in. in length at x100 magnification. The Roman numeral (see plate 1) indicates the graphite form and may not always be used if the type of iron is otherwise clearly defined.

Table I. Dimension of the Graphite Particles Forms I to VI.

Reference Number	Dimension of the Particles observed at x100 (mm)	True Dimension (mm)
1	>100	>1
2	50 −100	0·5 −1
3	25 − 50	0·25 −0·5
4	12 − 25	0·12 −0·25
5	6 − 12	0·06 −0·12
6	3 − 6	0·03 −0·06
7	1·5− 3	0·015−0·03
8	< 1·5	<0·015

Plate 1. Reference diagram* for the graphite form (Distribution A).

Form: I

II

III

IV

V

VI

*The diagrams show only the outlines and not the structure of the graphite.

Plate 2. Reference diagram* for the distribution of graphite (Form I).

Distribution: A

B

C

D

E

*The diagrams show only the outlines and not the structure of the graphite.

Plate 3. Reference diagrams* for the size of graphite flakes.

Size 1. Longest flakes 4 in. or more in length.

Size 2. Longest flakes 2—4 in. in length.

Size 3. Longest dimension 1—2 in. length.

Size 4. Longest dimension ½—1 in. length.

*The diagrams show only the outlines and not the structure of the graphite.

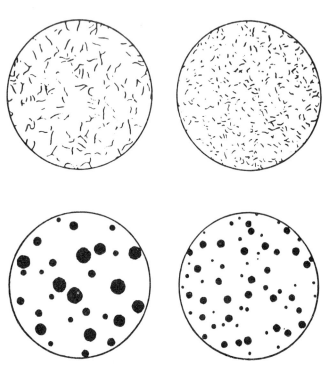

Size 5. Longest dimension
¼–½ in. length.

Size 6. Longest dimension
⅛–¼ in. length.

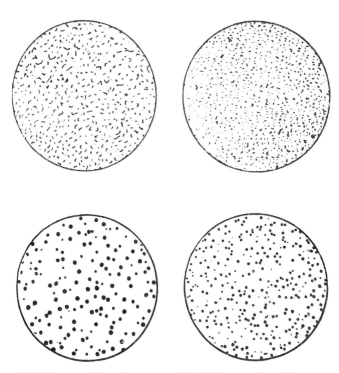

Size 7. Longest dimension
$\frac{1}{16}$–$\frac{1}{8}$ in. length.

Size 8. Longest dimension
$\frac{1}{16}$ in. or less.

MALLEABLE CAST IRON

Although a number of types of malleable cast iron have been developed, there is one feature common to all. The castings must possess an all white, graphite free structure as cast. Then an annealing heat treatment is applied in order to decompose the hard, brittle carbide and to develop the properties of malleability and ductility from which the product derives its name.

Types of Malleable Cast Iron

There are two main kinds known as whiteheart and blackheart. The names have reference to the colour of the fracture of the annealed material. Special malleable cast irons are modifications of these two original types.

Chemical Analysis

Castings for whiteheart malleable are produced mainly from cupola melted charges. The chemical analysis varies with the size and kind of casting between the following limits :—

Average limits of chemical composition for Whiteheart Malleable

	Before Annealing	After Annealing
Total carbon	3·0-3·7%	0·5-2·0%
Silicon ...	0·4-0·8%	0·4-0·7%
Manganese ...	0·1-0·4%	0·1-0·4%
Sulphur ...	0·1-0·3%	0·1—0·3%
Phosphorus	0·1% max.	0·1% max.

Iron for blackheart castings is melted in air, rotary or electric furnaces, although cupola melting is also performed. Duplexing, by first melting in a cupola followed by refining and adjustment of composition in some kind of hearth furnace is also carried out. Normal limits of chemical composition :—

Average limits of chemical composition for Blackheart Malleable before Annealing

	Air Furnace	Cupola
Total carbon	2·2-2·8%	3·0-3·3%
Silicon	0·8-0·95%	0·73-0·8%
Manganese	0·2-0·5%	2 x sulphur %
Sulphur	0·2% max	0·2% max.
Phosphorus	0·12% max.	0·12% max.

Average limits of chemical composition for Blackheart Malleable after Annealing

	Air Furnace	Cupola
Total carbon	2·1-2·4%	2·5-3·0%
Combined carbon ...	0·1%	0·1%
Silicon	0·8-0·95%	0·7-0·8%
Manganese	0·2-0·5%	2 x sulphur %
Sulphur	0·2% max	0·2% max.
Phosphorus	0·12% max.	0·12%max.

The compositions given here are average values and intended only as a guide. In practice the limits may vary somewhat more widely.

Mechanical properties of malleable cast iron

Three British Standard Specifications cover the three main malleable iron groups as follows:—

BS 310: 1972 Blackheart malleable iron castings
BS 309: 1972 Whiteheart malleable iron castings
BS 3333: 1972 Pearlitic malleable iron castings

Blackheart Malleable Iron

Because of the method of manufacture these irons have quite uniform and consistent properties underneath the immediate de-carburised casting surface. The matrix is fully ferritic and they are the most ductile of all the malleable irons. Elongation values are high as are the notched impact properties although the latter form no part of the specification.

Whitehart Malleable Iron

The microstructure of whiteheart malleable iron is variable through the section with a progressive increase in carbon content (present either as graphite or combined as pearlite) towards the centre. Mechanical properties will therefore be somewhat variable but good average results can be obtained consistently with good practice.

Pearlitic Malleable Iron

Of relatively recent introduction these irons are obtained by interrupting a normal blackheart anneal and air or oil quenching, or by heat treating and quenching subsequent to a normal blackheart anneal. The retained pearlitic matrix can also be obtained by varying the composition so that the required structure persists after a normal annealing cycle. These pearlitic malleable irons will give the highest mechanical properties available from this type of cast iron.

Tensile Properties

The three BS specifications given at the head of this section are further sub-divided into individual grades of iron as follows:—

BS 310: 1972 covering blackheart malleable irons includes grades B290/6, B310/10 and B340/12
BS 309: 1972 covering whiteheart malleable irons includes grades W340/3 and W410/4
BS 3333: 1972 covering pearlitic malleable irons includes grades P440/7, P510/4, P540/5, P540/5, P570/3 and P690/2

It will be noted that the prefix letter (B, W or P) indicates the type of malleable iron while the subsequent numbers indicate the minimum tensile requirement in Newtons/mm^2 followed by the elongation.

MINIMUM MECHANICAL PROPERTIES OF MALLEABLE CAST IRONS

1. BS.310:1972 Blackheart Malleable Iron

Grade	B290/6	B310/10	B340/12
Tensile Strength N/mm²	290	310	340
0:5% Proof Stress N/mm²	170	190	200
Elongation %	6	10	12
Hardness Brinell HB	124 to 140	127 to 140	132 to 140
Coef. of Thermal Exp. 10^{-6}/K	12.5	12.5	12.5
Specific Heat 20°C to 700°C J/kgK	665	665	665
Specific Gravity	7.35	7.35	7.35

2. BS.309:1972 Whiteheart Malleable Iron

Grade	W 340/3*			W 410/4*		
	9mm	12mm	15mm	9mm	12mm	15mm
Tensile Strength N/mm²	270	310	340	350	390	410
0:5% Proof Stress N/mm²	—	—	—	190	230	250
Elongation %	7	4	3	10	6	4
Hardness Brinell HB	Can vary across section. Values from 120 (surface) to 180 (centre) are generally obtained.					
Coef. of Thermal Exp. 10^{-6}/K	—12.5—			—12.5—		
Specific Heat 20°C to 700°C J/kgK	—665—			—665—		
Specific Gravity	—7.4—			—7.4—		

*Mechanical properties of whiteheart malleable irons will be affected by normal structural variations and it is therefore important that results are related to the average section thickness of the casting represented.

3. BS.3333:1972 Pearlitic Malleable Iron

Grade	P440/7		P510/4		P540/5		P570/3		P690/2
	Cooled in still air	Air blast	Cooled in still air	Air blast	Cooled in still air	Air blast	Cooled in still air	Oil Quench	Oil Quench
Tensile Strength N/mm²	440	440	510	510	540	540	570	570	690
0.5 Proof Stress N/mm²	290	293	318	352	330	377	402	433	577
Elongation %	7 to 11		4 to 9		5 to 8		3 to 7		2 to 3
Hardness Brinell HB	142 to 169		169 to 181		181 to 192		192 to 238		238 to 270
Coef. of Thermal Exp. 10^{-6}/K	12.5		12.5		12.5		12.5		12.5
Specific Heat 20°C to 700°C J/kgK	665		665		665		665		665
Specific Gravity	7.3		7.3		7.3		7.3		7.3

1 Newton per square millimetre (N/mm²) = 0.0645 tons-force per square inch (tonf/in²)
1 Ton-force per square inch (tonf/in²) = 15.4443 newtons per square millimetre (N/mm²)

NOTES ON BS. SPECIFICATIONS FOR MALLEABLE CAST IRON

As in the previous standards, minimum mechanical properties specified for each grade are determined using test bars representing the material quality of the castings. In addition to the use of metric units the following principal differences exist between these and the previous standards:

(a) BS. 309 and BS. 310 include a proof stress determination replacing a yield point measurement. Bend testing is not required.

(b) In all standards 0·5% proof stress values are required: the 0·2% proof stress values are given for information only.

(c) Brinell hardness values are included for guidance and are not part of the specification.

NODULAR CAST IRON
BS. 2789-1961

This British Standard now contains six grades of material instead of the three types specified in BS. 2789:1956.

TYPE SNG 24/17.—Those with a ferritic matrix in which resistance to impact is of paramount importance.

TYPE SNG 27/12.—Those with a mainly ferritic matrix of moderately high tensile strength in which high ductility and toughness are of great importance.

TYPE SNG 32/7.—Those with a ferritic-pearlitic matrix combining strength with reasonable ductility and forming an intermediate grade.

TYPES SNG 37/2, SNG 42/2 and SNG 47/2.—Those with a mainly pearlitic matrix, characterised by high tensile strength, but in which ductility and resistance to impact are of less importance.

Included in this revised specification is an optional impact test clause for Grade SNG 27/12 and a mandatory impact clause for Grade SNG 24/17.

An impact test for Grade SNG 27/12 is an optional requirement but where such a test is demanded, this grade should show an average impact value of not less than 6 ft./lb. for a set of three individual values and no individual test should show an impact value of less than 4 ft./lb.

The Brinell Hardness No. of Grade SNG 27/12 should not exceed 187, and that of Grade SNG 24/17, not to exceed 171.

Mechanical Properties of Nodular Cast Iron
British Standard 2789 : 1961

Grade	Tensile Strength Tons/sq. in. (min.)	0·5% Permanent Set Stress Tons/sq. in. (min.)	Elongation (% min.)	Impact Value ft. lb. (min).	
				Average*	Individual
SNG 24/17	24·0	15·0	17	10	9
SNG 27/12	27·0	18·0	12	6†	4†
SNG 32/7	32·0	22·0	7	–	–
SNG 37/2	37·0	25·0	2	–	–
SNG 42/2	42·0	28·0	2	–	–
SNG 47/2	47·0	30·0	2	–	–

*Average of set of three on one sample.

†Optional.

Nodular cast iron or Spheroidal Graphite (S.G.) iron may be defined as a cast iron having a microstructure containing at least 90% of the graphite in the truly spheroidal form. Several elements are known to be capable of developing this type of structure, magnesium and cerium being the elements most widely used for this purpose. Other elements however, such as calcium and lithium, are also known to develop nodular type structures.

THE MAGNESIUM PROCESS

Of the elements known to be capable of developing nodular graphite structures when added to molten cast iron, magnesium offers the best industrial interest and seems to be able to develop the best mechanical properties. A further advantage of the magnesium process over the cerium, is the somewhat wider range of base iron composition permissible with it.

Composition of Base Iron

Sulphur must be of a low order in the iron to be treated. Magnesium is also an active desulphurising agent and its first effect is to remove sulphur down to a level of about 0·01%. Until the sulphur is reduced to something like this limit, the magnesium exerts little influence on the graphite formation. In the formation of magnesium sulphide, 0·1% sulphur requires 0·075% magnesium.

Apart from the sulphur aspect, the preferred composition of the base iron is within the following limits:

Total carbon	3·4-4·2%
Silicon	1·8-2·5%
Manganese	0·4% maximum
Phosphorus	0·2% maximum

Whilst manganese and phosphorus do not prevent the formation of graphite nodules, they do reduce ductility very considerably and since this property is an important feature of nodular cast iron, it is essential to maintain these two elements at as low a level as possible.

Introduction of the Magnesium

One of the greatest practical difficulties is getting the required amount of magnesium into the melt with the necessary degree of consistency and assurance, combined with freedom from risk of danger. Magnesium boils at about 1120°C, and when plunged into molten cast iron at 1400°C, magnesium metal melts and vaporises instantaneously, escaping with violence and carrying some of the cast iron with it.

Successful magnesium treatment requires an emulsion of magnesium vapour and liquid cast iron. To obtain an intimate emulsion, liberation of the magnesium should occur over a period of time and at many places in the molten iron. This can be accomplished by using alloys of magnesium with one or more elements which are soluble in iron or a specially made impregnate such as MAG-COKE.

MAG-COKE

MAG-COKE is a nodularising agent consisting of metallurgical coke impregnated with metallic magnesium. It contains approximately 43% by weight of the metal and 57% of coke. The coke acts as a carrier for the magnesium and also moderates the reaction so that when MAG-COKE is plunged into molten iron the release of magnesium vapour is controlled and steady.

Because only magnesium is added to the metal certain advantages follow:

1. There is a greater flexibility of charge materials since there is no build-up of constituents such as silicon (from the use of magnesium ferro-silicon).

2. The attractive economics mean that desulphurising and nodularising can be carried out in one operation. Separate treatments pose problems of separate materials and equipment (porous plug ladles, shaking ladles, etc.)
3. Possible difficulties with high aluminium contents (added with magnesium ferro-silicon) and excessive slag or dross are avoided.

MAG-COKE is best added to the metal by plunging since its low density usually precludes "sandwich" or "pour-over" methods. The efficiency of magnesium recovery depends on a number of factors—temperature, ladle size, method of use, sulphur content of iron, required residual magnesium, etc. The following equation may be used as a guide to indicate the order of recovery to be expected:—

$$\text{Recovery \%} = \frac{\text{Retained magnesium}}{\text{Added magnesium} - \frac{3}{4}\text{ base sulphur}} \times 100$$

For a treatment weight of 10 cwts. and a base sulphur level of 0·01% or less, magnesium recovery will generally fall within the range 25-50%. The lower figure will obtain at around 1500°C metal temperature and the higher at 1400°C using good treatment practice. Increased melt weight gives higher recoveries. Temperature loss when treating 1 ton of iron with 0·5% MAG-COKE will be of the order of 30-35°C only, provided that the plunger, etc., is strongly preheated.

Inhibiting Elements

Certain constituents or elements that may be present in cast iron have an inhibiting effect on nodule formation. Experimental work on approximately 1 in. thick castings indicates that the following elements in the amounts shown have an inhibiting effect:

Aluminium	above 0·13%
Arsenic	above 0·09%
Bismuth	above 0·002%
Lead	above 0·005% completely suppresses formation of nodules
Tin	above 0·04%
Titanium	above 0·04%

To this list can also be added antimony, tellurium and selenium although the limiting figures are not known.

All these exert a detrimental effect on the development of the nodular graphite structure. Also the combined effect of these elements is at least additive and may be even more potent. So when

276

two or more are present in amounts which would not individually affect the structure, they may together prevent the formation of the nodular graphite. Some of the elements which cause interference are also capable of producing limited nodularisation, namely, bismuth, selenium, tellurium and titanium. This anomaly suggests that nodularisation may take place by two or more different mechanisms. Cerium added to a magnesium treated iron will neutralise the effect of interfering elements.

Quantity of Magnesium Required

During the magnesium treatment, part of the element is lost by evaporation. Part oxidises and reacts with the molten iron, forming compounds, both soluble and insoluble, some of which enter the slag. The chief slag forming reaction is that producing magnesium sulphide. It is only the residual magnesium which remains in the cast iron which is effective in promoting nodularisation of the graphite.

The residual magnesium in nodular cast iron is only a few hundredths of 1%. Limiting values which have been noted are 0·025% to 0·12%. C. K. Donoho considers that the best mechanical properties are obtained with a residual magnesium content of about 0·06%.

Like cerium, magnesium exerts a carbide stabilising action and when present much beyond 0·1% it is likely to give white or mottled structures, particularly where thin sections and rapid cooling are concerned.

Efficiency of magnesium recovery varies considerably but it seems to be greatest in the case of MAG-COKE and the nickel and copper alloys carrying 20% or less of magnesium. Recoveries reported vary from 25% up to over 60%. A magnesium ferro-silicon containing approximately 7% magnesium, approximately 45% silicon, the balance mainly iron showed the following recoveries:

Pouring temperature	Magnesium recovery
1454°C.	35%
1399°C.	42·6%
1343°C.	65·0%

Inoculation Treatment

Immediately following the addition of the magnesium bearing alloy, the melt has to be inoculated with from 0·5% to 0·75% of a graphitising inoculant such as INOCULIN 1. This is to offset the carbide stabilising action of the magnesium which operates in a similar manner to that of cerium.

Some claims have been made that when using the high silicon magnesium alloys such as magnesium ferro-silicon, no further inoculation is required. Whilst some corroboration of this has been obtained, it is not possible to be quite positive one way or the other in the light of present knowledge.

Influence of Other Elements

As with the cerium process, nickel, copper and molybdenum may be present in any proportions, either separately or combined. In this way, special austenitic, martensitic or acicular matrix structure can be obtained.

On the other hand, there are a number of elements which are known to inhibit development of the graphite nodules. These have already been mentioned and how their deleterious effects can be overcome by the combined use of magnesium and cerium.

Heat Treatment of Nodular Cast Irons

Successful treatment with cerium results in graphite nodules in a mixed matrix of ferrite and pearlite. Tensile strength is about 35 ton per sq. in., with 2-4% elongation.

The magnesium irons in the "as cast" state, have the graphite nodules in a matrix more predominately pearlitic. Consequently tensile strength is higher, up to 45 ton per sq. in., with less elongation, at about 2%.

By comparatively short annealing treatment at 850-900°C for a time which varies with composition and section thickness, all nodular irons can be converted to the all ferritic matrix. In this condition they resemble blackheart malleable in structure and will record a tensile strength of about 30 ton per sq. in. and elongation of 10%.

Thus there are three types of nodular cast iron available.

1. The cerium iron, mainly ferritic as cast.
2. The magnesium iron, mainly pearlitic as cast.
3. Either of these rendered fully ferritic by annealing. All have some degree of ductility, and impact values greatly exceeding those of any form of grey cast iron hitherto produced.

THE CERIUM PROCESS

The Cerium Addition

Cerium may be added to cast iron as a wide variety of alloys. Pure cerium, iron-cerium, nickel-cerium, copper-cerium, silicon-cerium, manganese-cerium, and aluminium-cerium all dissolve quite easily in molten cast iron. On account of its ready commercial availability and its relatively low cost, a cerium alloy known as

278

mischmetall is most frequently employed. Mischmetall contains approximately 50% of cerium, the balance being largely other rare earth elements, the most important of which is lanthanum. A typical analysis is as follows:

Cerium	45-53%
Lanthanum	22-25%
Neodymium	15-17%
Other rare earth elements	8-10%
Iron	Up to 5%

It is a relatively low melting point alloy of a high density, fairly soft and somewhat brittle, but it cannot be crushed easily into small pieces. A hack-saw may be used to cut it, but sparking will occur due to the pyrophoric nature of the alloy. (Cerium is an important constituent in lighter flints.)

For adding to molten cast iron, the separate pieces of mischmetall should not weigh less than about 20 grams, or more than about 300 grams. The important point to be observed is that mischmetall should not be finely divided in the form of powder, as loss by oxidation may occur. On the other hand, the pieces should not be so big that difficulty in obtaining uniform solution is experienced.

Influence of Cerium

Cerium is a powerful desulphuriser. When the sulphur content of a cast iron exceeds about 0·02%, the effect of adding cerium is first to reduce the sulphur content. Cerium combines preferentially with sulphur even in the presence of manganese and the cerium sulphide so formed rises to the surface of the molten metal.

When the sulphur content of the iron is 0·02% or less, the cerium will function as a carbide stabiliser and will give nodular graphite structures under the appropriate conditions. As a general rule cerium in cast iron will not function as a carbide stabiliser and will not give nodular graphite unless the final analysis of the solidified casting shows less than about 0·02% sulphur and more than about 0·02% cerium.

The higher the sulphur content of the molten iron into which the cerium is added, the greater will be the amount of cerium required. As cerium is a relatively expensive addition in any form, it is obviously of great importance to limit the amount of the element required by maintaining the sulphur content of the iron to be treated, at the lowest possible level.

Composition of the Iron to be Treated

Base iron composition must be such as will solidify grey before treatment. The following points should be observed:

279

1. TOTAL CARBON CONTENT. This should be such that when the percentages of silicon and phosphorus are added together and divided by 3 and the result added to the total carbon content, the value resulting should be more than 4·3%.
2. SILICON CONTENT. After the treatment with cerium and the inoculant, silicon should be in excess of 2·3%, but not greater than 3·3%. When nickel and copper are present, the minimum silicon may be slightly less, but not lower than 1·8%, provided sufficient nickel and/or copper is present to give an equivalent silicon content of 2·3%, assuming 1% silicon equals 3% of nickel or copper.
3. MANGANESE CONTENT. The manganese content can have any value. However, apart from austenitic compositions it should be kept below 0·5% mainly in the interest of obtaining maximum ductility. The low sulphur of the treated iron allows the manganese present to exert its full carbide stabilising influence.
4. PHOSPHORUS CONTENT. Phosphorus has an embrittling effect on nodular cast irons. On no account should it exceed 0·5% for in excess of this the solubility of the cerium is reduced to a figure below the minimum required for the formation of nodular graphite. For development of the best mechanical properties the phosphorus should be kept at 0·1% maximum, or less.
5. SULPHUR CONTENT. The importance of keeping this low has already been stressed. After treatment with cerium it should be below 0·02%.

Control of Sulphur Content

With crucible or electric furnaces, sulphur content can be kept low by the selection of charge materials containing little of this element. Hematite pig irons can be obtained with sulphur contents below 0·03%. When such melting stock is used, no preliminary desulphurising treatment need be applied.

Although a similar technique can be applied to cupola melting, there is always some sulphur pick-up from the coke. Thus, in most cases, a desulphurising treatment will be necessary before making the cerium addition.

Addition of Cerium and Inoculant

Cerium in the form of mischmetall dissolves very readily in molten cast iron. It may be added to the molten metal in the ladle and then plunged or stirred in with a refractory plunger. Alternatively, it may be placed on the bottom of the ladle and base iron of correct composition teemed on to it. In most cases the "double treatment" technique will be adopted. This involves the use of a graphitising inoculant added either simultaneously with, or preferably immediately following, the cerium addition. INOCULIN 1 is recommended in a ratio of from 0·5% to 0·75% of the weight of the iron being treated.

When low sulphur or desulphurised iron is tapped direct on to the mischmetall, the inoculant can be added to the surface of the molten metal and stirred in. In many cases, it may be more convenient to place both the inoculant and the mischmetall in the bottom of the ladle, so that both are added simultaneously and stirring of the latter is not required. This method gives uniform results and good recovery of cerium, but the inoculant must not be added before the cerium. As the mischmetall probably dissolves more rapidly than the inoculant, the correct order of solution may be obtained in this way.

Cerium, as pure cerium, mischmetall, or any of the cerium alloys mentioned, dissolves in molten cast iron without any violent reaction, and the process is not hazardous in any way.

Scum Formation

Following the addition of the mischmetall and the inoculant, a scum of cerium sulphide tends to form on the surface of the melt. The quantity is dependent on the initial sulphur content. Since this slag is usually formed, precautions should be taken to prevent it entering the castings. Resort can be made to careful skimming, teapot spout ladles, the use of slag trap runners, and slag coagulants specifically developed for S.G. iron.

Metal Temperature

When adding the cerium and inoculant, metal temperature should be between 1350° and 1500°C. When cerium is added to hyper-eutectic irons at temperatures below 1300°C there is a danger that some flake graphite will already have formed in the melt before the addition is made. Any flakes so formed will be unaffected by the cerium addition and will subsequently cause inferior mechanical properties.

After making the addition any convenient pouring temperature may be used.

Residual Cerium

When the sulphur content of the metal is below 0·02% in the base iron, the amount of cerium found by analysis in the solidified casting should be between 0·02% and 0·1%. For most purposes, it is sufficient to have between 0·03% and 0·06% of cerium. If the initial sulphur content is higher than 0·02%, too much of the cerium may be expended in forming cerium sulphide and the nodular structure will not be obtained.

Cerium dissolved in cast iron has a powerful carbide stabilising effect, but when the cerium treatment is followed by the addition of graphitising inoculant this effect is largely neutralised. Provided the cerium content falls within the range 0·03% to 0·06% accurate control is not necessary. For thin sections of 5 mm or less, it is advisable to keep the cerium at 0·04% maximum.

Quantity of Cerium

In most cases the correct amount of cerium to be added must be found by experiment. It will vary according to the method of melting and the sulphur content of the iron to be treated. The following examples serve to indicate the range of additions necessary when using mischmetall containing about 50% of cerium.

1. A hematite pig iron having 0·015-0·025% sulphur when crucible melted will require a minimum addition of mischmetall equivalent to 0·15-0·25% of the weight of the metal being treated.
2. Cupola melted hematite iron not desulphurised, with a sulphur content of 0·06% will need a mischmetall addition equivalent to 0·45-0·75% of the weight of the metal being treated.
3. Cupola melted hematite iron desulphurised, and having, after desulphurising, a sulphur content of 0·03%-0·04% will require an addition of mischmetall equivalent to 0·3-0·5% of the weight of the metal to be treated.

Material which exhibits a predominantly flake graphite structure after treatment indicates a deficiency of cerium, and more mischmetall should be added. On the other hand a white or mottled fracture—except in the case of chilled castings—shows that the amount of mischmetall added needs to be reduced.

SCRAP DIAGNOSIS — ITS CAUSE AND CURE
with particular application to Iron Castings

Many foundrymen are often in doubt when faced with a scrap casting, and very often the cure is obtained by rule-of-thumb methods, no direct approach to the trouble being made. It is hoped that the following list will help in a quicker diagnosis of scrap and its cure.

DEFECTS	APPEARANCE	CAUSE	REMEDIES
MISRUN	May appear as : 1. Holes in the thin sections of a casting. Edges are smooth and well rounded ; surface of metal round holes smooth and often shiny. 2. As a line, when two streams of metal have met but not fused together. Fracture may occur along this line. Casting surface may be smooth and shiny.	1. Low pouring temperature. 2. Low fluidity. 3. Hard mould (mainly on very thin casting). 4. Core shift, causing uneven thicknesses. 5. Vent low on sand. 6. Very high moisture content. 7. Pouring practice.	1. Provide hotter metal at cupola spout ; reduce heat losses in ladle by using flux coverings. 2. Increase Carbon and Phosphorus. 3. Avoid excessive ramming. 4. Take extra care in positioning. 5. Increase vent by means of vent wire or by adding Silica sand in mixing. 6. Reduce moisture compatible with moulding. 7. Keep runner bush full of metal during pouring.
SHRINKAGE and DRAWS	Rough cavities entering casting on heavy sections, or at the joint of change of sections. Saucer-shaped depressions on heavy sections, usually with rough edges.	1. Incorrect gating and feeding.	1. Use risers to feed heavy sections and ensure that they are filled with hot metal. If using open risers, use feeding flux, if using blind risers use feeding cores. Embody chills where a heavy section or boss cannot be fed directly with a riser. (See also pages 81 to 86.)

Scrap Diagnosis — Its Cause and Cure—*continued*

DEFECTS	APPEARANCE	CAUSE	REMEDIES
SLAG	Similar to above. See Shrinkage and Draws (previous page). Cavities are generally more saucer-shaped and smoother. Slag may be seen before cleaning the castings.	1. Dirty metal. 2. Incorrect gating.	1. Remove all slag from metal before pouring. Thicken slag with sand before skimming. 2. Incorporate skim gates or strainer cores in runner systems. Keep runner bush full whilst pouring.
POROSITY	Castings "weep" under pressure test. Machined surfaces show cavities in thick sections or a series of pin-holes on machined skin.	1. Wrong type of metal. 2. Running and feeding system. 3. Gassy metal.	1. Reduce Silicon or Phosphorus content. 2. See shrinkage. 3. Degas and scavenge well.
HARD METAL	Bright areas on machined faces, often at corners or edges of thinnest sections. May occur as scattered hard spots. Shows, when broken, a white fracture.	1. Wrong type of metal. 2. High moisture content. 3. Pouring practice.	1. Increase Silicon content by : (a) altering metal mixture ; (b) introducing Silicon into ladle. 2. Reduce moisture. 3. Avoid splashing metal down runners and risers — "plug" sprues very helpful.
SCABS	Rough "warty" excretions on surface of casting, mainly on heavy sections.	1. Uneven ramming. 2. Incorrect gating. 3. Improperly dried moulds. 4. High clay content in moulding sand.	1. Ram more evenly. 2. Gate so that an even flow of metal is obtained over surface of mould. 3. Avoid too rapid drying and allow time for heat to penetrate through the mould. 4. Change moulding sand.

Scrap Diagnosis — Its Cause and Cure—continued

DEFECTS	APPEARANCE	CAUSE	REMEDIES
DISTORTION	Casting shows swelling on surface.	1. Uneven mould hardness due to insufficient ramming to withstand metal pressure. 2. Poor weighting practice.	1. Ram evenly and firmly. 2. Increase weight on moulds and ensure it is distributed evenly.
ROUGH SURFACE	Casting surface rough.	1. Metal penetration. 2. Low coal dust content. 3. Hard sand surface with low permeability. 4. Moulding sand too "open."	1. Use finer sand or use mould dressing. 2. Make additions of coal dust. 3. Open up the sand. 4. Close-up sand somewhat.
CRACKS	Hair line cracks showing on casting. When broken, discolouration shows that crack was produced while casting was hot. No discolouration shows cold crack.	1. High dry strength of sand. 2. Cores too hard. 3. Casting strains. 4. Mechanical reasons.	1. Ram softer to allow casting to contract. 2. Reduce oil in cores. 3. Gate evenly to avoid these. Break mould to allow of free contraction. 4. Pack casting with wood or old tyres in tumbler. Take care in breaking off risers. See that risers are provided with correctly designed necks.

Scrap Diagnosis — Its Cause and Cure—continued

DEFECTS	APPEARANCE	CAUSE	REMEDIES
BLOWHOLES	Rough shaped holes occurring on the outside of the casting or in the thicker sections. May be found just below surface on machining. In severe cases, section of casting may be hollow. Cavities may be dull or bright depending on conditions under which they have formed.	1. Low vent on moulding or core sand. 2. Hard ramming. 3. High moisture content. 4. Rusty or damp chills and chaplets. 5. Very hard cores. 6. Insufficient venting in cores. 7. Incomplete baking. 8. Damp pouring ladles. 9. Too low a pouring temperature.	1. Increase vent by use of vent wire or open sand with additions of Silica or by the use of a coarser Silica sand. 2. Avoid excess ramming. 3. Reduce moisture to minimum, consistent with workability. 4. Ensure chaplets are dry, and coat chills with oil or proprietary dressing before use. 5. Reduce oil in sand. 6. Ensure vents are clear. 7. Bake until centre is dry and hard. 8. Thoroughly dry all pouring ladles. 9. Increase pouring temperature.
DIRT	Rough cavities and pits in casting surface. If examined before cleaning the sand may often be seen.	1. Strength of sand low. 2. Loose ramming. 3. Direct wash of metal on sand surface, e.g., cods, corners, etc. 4. Poor finish of gating system. 5. Displacement of sand by cores. 6. Disturbed moulds. 7. Insufficient taper on patterns.	1. Increase green bond. 2. Ram evenly. 3. Avoid direct wash with well designed runners. 4. Finish of running system should be as good as mould. Make bushes and runners with good facing sand. 5. Blow out after placing cores. 6. Place weights carefully ; avoid knocking moulds. 7. Increase taper to allow clean lift.

REMOVAL OF SAND FROM IRON CASTINGS BY PICKLING

The pickling solution is made up from water and a mixture of equal parts of hydrofluoric and sulphuric acids. The mixed acids being present to the extent of 10%. A lead-lined tank is required for holding the pickling solution. For preference a match board lining is fitted in order to protect the soft lead. The size of the tank is determined by the size of the casting and the output required, allowing about four hours for pickling.

Two other tanks are required, one to contain clean running water, and the second an iron tank heated either by steam coil or gas burner, and filled with a dilute solution of soda.

To start up, No. 1 tank is filled to just below the desired working height with water, and hydrofluoric and sulphuric acids are added at the rate of one litre of each to every 20 litres of water. (Always add acid to water to avoid accident—the concentrated acids are dangerous and skin contact, splashing, etc., must be avoided.)

No. 2 tank is filled with water, and No. 3 tank with a 10% solution of washing soda, and heated to near boiling.

Suspend the castings in No. 1 tank by means of mild steel or Monel metal hooks and allow to remain a sufficient time to remove or soften the sand and core adhering to the casting. The length of time will be judged by inspection, the usual period being between three and four hours. Remove the castings and swill with a strong jet of water to remove all adhering sand and pass through the water and soda tanks. Leave in the last tank long enough for the casting to get nicely warm, so that penetration of the neutralizing soda is obtained and sufficient heat absorbed to dry the castings quickly. Finally rinse with clean water.

The strength of the acid bath is maintained by daily additions of an equal quantity of each acid, the requirements are judged by the evolution of gas from the castings when in pickle, which should resemble the rise of gas from soda water recently drawn from a syphon.

As a rough guide it should be possible to pickle an average of 30 four-cylinder motor engine blocks with 5 litres of each acid.

Mud and insoluble dirt accumulate in the pickling tank and must be cleaned out from time to time. The period varies with amount of castings put through ; on an average with a pickling tank going continuously, cleaning out will be needed at the end of every six to eight weeks.

Measurement and handling of the acids are performed by means of a lead jug, and the hands of the operators are protected by gauntlet rubber gloves.

The pickling process described removes all traces of sand and other siliceous material. Not only does it produce a cleaner looking casting but also effects considerable economies in subsequent machining operations by cutting down tool costs.

Waste material from the process including sludge from the bottom of the tanks as well as "killed" acid must be disposed of very carefully as it is often extremely poisonous and corrosive.

CAST IRONS FOR RESISTING HEAT

CLASSIFICATION OF HEAT-RESISTING CAST IRONS

These may be grouped under five main classifications :—

1. Certain white and unmachinable cast irons which do not grow under the influence of high temperature.

2. Close-grained cast irons containing low silicon and carbon contents, and small quantities of phosphorus, corresponding with good engineering properties, and sometimes having alloy additions of the order of about 3%.

3. Irons of the Silal type with silicon in excess of 4%.

4. Austenitic cast irons.

5. High chromium Ferritic cast irons.

I. White Cast Irons

The carbon in white cast irons is all in the combined form and there are no graphite flakes. When the composition is such that the combined carbon is stable at high temperatures, no growth results when the iron is repeatedly heated and cooled, but there is a tendency to develop cracks because of its brittleness. An iron with the following composition was found to be immune from growth at 900° C. : Total carbon 2·66%, silicon 0·59%, and manganese 1·64%. Although white irons do not grow, they scale rapidly under oxidising conditions, and their application is limited because of their brittleness and unmachinability.

2. Low Silicon Irons

Such irons, with about 1·2% silicon, and about 3·2% total carbon, have a fine, close-grained structure. They are also strong and possess some ductility at normal temperature. Owing to the "critical" or phase change demarcation which takes place between 700° and 750° C., it is apparent this class of iron should not be utilised for components whose normal service life necessitates working at temperatures above a maxi..um of 700° C. Service, at temperatures up to that mentioned, from this grade of cast iron can be improved by the addition of chromium up to 0·5% and nickel at 1·5 to 2·0%.

Irons under this heading have the advantage over those in Groups 3 and 4 in that they are easier to make into sound castings which will withstand rough usage. On the other hand, they oxidise readily, so where oxidising conditions are severe, they fail quickly by scaling.

F —K

Small changes in the amounts of sulphur, manganese, and phosphorus, have no appreciable effect on the tendency of these irons to grow at high temperatures. The amount of phosphorus which can be allowed depends upon the conditions. Below 900° C. phosphorus helps to maintain the iron rigid. Above this temperature it rapidly becomes plastic and the iron may collapse. The lower the phosphorus the greater the ductility and hence the less liability to cracking through repeated heating. Thus for use at temperatures below 900° C., up to 0·75% phosphorus may be present, but for castings subjected to more than 900° C., the phosphorus content should be about 0·3% maximum.

3. High Silicon Irons

Developed by the B.C.I.R.A. and covered by British Patent No. 323,076, known by the registered trade mark "Silal." The group carries silicon contents ranging between 4% and 10% and consequently possesses a structure of finely distributed graphite in a ferritic matrix. Their resistance to high temperature deterioration is due to the following reasons :—

1. Low total carbon content, yielding a small graphite quantity and structure, which ensures the maximum resistance to gas penetration.

2. Resistance to oxidation by reason of high silicon content.

3. The volume change points can be raised beyond the maximum service temperature by the use of sufficient silicon.

In any application, the actual silicon content is determined by the temperature ; the higher the temperature, the higher the silicon content required. As the silicon is increased, the irons become weaker and more brittle at normal temperatures and it follows therefore that the minimum silicon content necessary to prevent growth should always be used. In the ordinary way, the lowest silicon contents vary between 5% and 6%. This grade, however, is not recommended for service temperatures above 900° C. Above that temperature, scaling and growth take place, but will not occur if irons of higher silicon content, up to 10%, are used. Total carbon should be about 2·5% at less than 0·5%. Manganese should not exceed the quantity necessary to neutralise the sulphur present, that is, with 0·1% sulphur, the manganese should be 0·5%. An iron such as this can be made readily in the cupola, from a charge composed of ferro-silicon and steel scrap, because the solubility of carbon in molten cast iron is lowered by 0·25% by each 1% of silicon present.

4. Austenitic Irons

To overcome the brittleness of "Silal" attempts were made to impart some ductility by adding nickel and chromium. The resulting product was called "Nicrosilal," British Patent No. 378,508. It is a grey, non-magnetic cast iron, containing for most purposes, total carbon 2·0%, silicon 5·0%, manganese 1·0%, nickel 18%, chromium 2 to 5%. It is tough and ductile and resists growth up to 1,000° C.

No difficulty is experienced in getting a total carbon as low as 2% from a cupola melt, because the solubility of carbon in molten cast iron at a given temperature is reduced about 1·25% by 5% silicon and about 0·7% by 18% nickel. The lower the total carbon content, the greater is the strength and heat resistance, but the greater also is the liquid shrinkage and the tendency to mottle in the thin sections.

For most purposes a silicon content of 5% is best. With more than 5% silicon and less than 17% nickel, the irons harden on tempering at 600° C. and when this occurs, the volume increases and the iron becomes magnetic.

Although 18% nickel is necessary to render the iron austenitic when no chromium is present, this amount is decreased to 13% as the chromium rises to 1·8%. Hence, the minimum content is 13%, and the maximum may be as high as 20%.

Manganese content should not be above 1%, and phosphorus should be below 1% for temperatures up to 950° C.

"Ni-resist" is another austenitic iron of composition :

	Min.		Max.
Total carbon ...	2·7%	to	3·2%
Silicon	1·0%	,,	2·0%
Manganese	0·8%	,,	1·5%
Sulphur	—	,,	0·12%
Phosphorus	—	,,	0·40%
Nickel	12·0%	,,	16·0%
Copper	6·0%	,,	8·0%
Chromium	1·5%	,,	4·0%

For sections above 50 mm, in order to ensure free-machining, it is desirable to increase the nickel content by a further 2% to 3%.

It consists essentially of an alloy of cast iron with Monel metal, with further additions of such elements as chromium. A suitable composition for the base iron is : Total carbon 3·3 to 3·5%, silicon 1·4 to 2·25%, manganese 0·5 to 1·2%, sulphur 0·08% maximum, phosphorus not over 0·5%. The nickel and copper may be added either as a separate addition in any suitable form, or in the form of Monel ingots. A nickel-copper-chromium pig iron is produced which contains the special elements in the required proportions.

5. High Chromium Cast Irons

These also have useful application in the high temperature range where resistance to scaling is a major requirement. The structures of these irons consist of primary iron-chromium carbides in a matrix of iron-chromium solid solution and secondary iron-chromium carbides associated with ferrite. There are two main grades. The first contains 16% of chromium with about 1·9% each of carbon and silicon. It is produced under the registered name "Enduron" by the Sheepbridge Stokes Company. The Vickers Diamond hardness is between 280 and 320, which makes it more difficult to machine than the heat-resisting irons mentioned previously. Also it is more susceptible to cracking. However, it resists scaling at temperatures from 900° to 950° C.

An iron containing chromium at 33% with total carbon 1·2% and silicon 2·0% has the best scale resistance of any of the heat-resisting irons and can be used at temperatures as high as 1050° C. without marked deterioration. As in the case of the 16% chromium iron it is not easily machined and design of parts must be considered in relation to its increased brittleness and greater liability to cracking.

Summary

A general summation of the applicability of different kinds of cast iron compositions to service at high temperatures can be drawn up as follows :—

Temperatures up to 600° C. Straight unalloyed grey irons of the high duty type and possessing sound, dense, fine-grained structures will give satisfactory service up to about 600° C. In the lower temperature ranges phosphorus is not a harmful element so far as growth and scaling are concerned. Actually, in the amounts usually present in light casting mixtures it tends to improve scaling resistance under conditions of heat and oxidation within the limits of temperature stated. Phosphorus also imparts increased stiffness and rigidity, provided these temperatures are not exceeded.

Temperatures up to 700° C. Here again there is no real need to depart from grey iron compositions within the category encompassed by the close-grained alloys described in the group two classification. For this increased temperature range a simple ladle addition of chromium will confer the required stability of the pearlite and decrease the size of the graphite flakes.

Chromium is effective in retarding decomposition of the carbides at a ratio of about 0·6%. Above this value there is some risk of obtaining too great a hardness for easy machinability, especially in thinner sections. However, undue development of hardness can be guarded against and higher percentages of chromium used if a suitable proportion of silicon is introduced simultaneously with the chromium.

Where conditions of heat shock have to be withstood it is wise to use a relatively high total carbon content so as to exploit to the full the buffer or cushioning effect of the higher graphite precipitation in the presence of large temperature differentials. An example of a typical composition follows :—

Total carbon	...	...	3·40—3·70%.
Silicon ...	...	...	according to section of casting.
Manganese	...	...	about 1%.
Sulphur	...	...	less than 0·12%.
Phosphorus	...	...	0·3% maximum.
Chromium	...	...	0·6%.

The silicon content will vary according to the average sectional thickness of the casting :—

From 9 mm to 18 mm average section thickness 2·3% silicon.
From 18 mm to 38 mm average section thickness 2·1% silicon.
From 38 mm to 60 mm average section thickness 1·9% silicon.

Temperatures up to 850° C. Still keeping within the group two range of composition, a 1% addition of aluminium will reduce oxidation and scaling losses up to 850° C. At this temperature the aluminium increases the life to about five times that of unalloyed grey cast iron. A somewhat greater resistance to heat and scaling can be imparted by raising the aluminium from 2% to 4%. However, ratios between 1% and 2% must be avoided. With the higher aluminium contents, steps must be taken to guard against the inclusion of defects caused by the entrapment of aluminium oxide in the castings.

293

Temperatures up to 900° C. For this relatively high temperature resort must be made to irons of the high silicon silal type described in group three.

A composition which can be recommended is :—

Total carbon	...	...	...	2·2%.
Silicon	...	...	...	6·25%.
Manganese	...	...	...	0·50%.
Sulphur	...	...	...	lowest.
Phosphorus	...	...	...	0·50% maximum.
Copper	...	...	...	1·50%.
Chromium	...	...	...	0·50%.

The cupola charge for this is based on silvery pig iron or ferro-silicon and arranged to yield a silicon content of 6·5%. Copper and chromium are added to the molten metal in the ladle.

Temperatures above 900° C. Here it is necessary to make use of the highly alloyed austenitic compositions indicated in group four. Some of these have well known proprietary names such as Nicrosilal, Ni-Resist and Causal, which are the subject of patents.

Useful alternatives to the austenitic series are the class 5 Irons of 16% and 33% chromium content. Their resistance to scaling is very good and chromium is a less expensive alloying element than nickel.

CAST IRONS FOR RESISTING CORROSION

When considering corrosive attack the position is rendered much more difficult than for heat resistance because of the almost infinite combination of corrosive agents, their concentrations, temperature and other conditions. Therefore, every case must be considered on its own as an individual problem. It is possible, however, to give general guidance as to what type of composition of iron will be most likely to give satisfactory service in particular cases.

No attempt can be made to cover the action of all possible corrosive media but perhaps it is true to say that cast irons which can withstand attack by the three principal mineral acids, hydrochloric, nitric and sulphuric are likely to have a useful life under many other conditions of acid corrosion.

I. Low Alloy Cast Irons

These cannot be expected to show any outstanding improvement over the range of unalloyed cast irons, except in so far as small additions of alloying elements contribute towards stronger, sounder structures and freedom from non-metallic inclusions. An outstanding instance is the resistance against caustic alkali attack which is imparted to cast iron containing up to 2% of nickel and 0·5% of chromium.

2. High Nickel Cast Irons

The high nickel austenitic cast irons have been described in group 4 of the heat-resisting irons, page 291. Of these, the best known is Ni-Resist which has satisfactory founding properties, is readily machinable and considerably tougher than ordinary grey cast iron. Its tensile strength is not very high, about 200 N/mm^2.

Ni-Resist compositions withstand well the action of sulphuric acid at atmospheric and moderately raised temperatures in all strengths from about 20% to 80% and up to 60° C. Actually it performs better than 18/8 stainless steel and is probably the most economical material for standing up against sulphuric acid corrosion.

With hydrochloric acid Ni-Resist iron is not so good. In the case of cold dilute acid it has a field of usefulness, but is unsuitable in the case of hot and strong hydrochloric acid solutions.

For nitric acid Ni-Resist is not suitable at all, but it offers an excellent resistance to caustic soda over a wide range of strength and temperatures. Under conditions of attack such as with industrial atmospheres, sea and tap water, surface rusting occurs but the rate of corrosion falls off with time to about one-fifth to one-eighth that of ordinary grey cast iron.

Copper is normally present in Ni-Resist compositions and produces a blue corrosion product which cannot be tolerated in the food, soap and cosmetic industries. Austenitic irons without copper, and containing about 21% nickel or Nicro-Silal are then employed. Cast irons of this type are non-magnetic and have a higher coefficient of thermal expansion than that of ordinary iron.

3. High Chromium Cast Irons

Item 5 in the section dealing with heat resisting cast irons on page 292 gives a general description of these. Although they have a much higher tensile strength than the austenitic irons (550 N/mm^2 with 30 to 33% chromium) they are not so tough. Also their hardness of approximately 300 Brinell makes machining more difficult.

Because chromium has the property of forming continuous oxide films of high stability it is able to confer a high resistance to corrosion under strongly oxidising conditions such as are met in resisting the attack of nitric acid. To exert its maximum effect the chromium must be present in solid solution and not in the form of carbide. Thus with chromium at 33% and just over 1% of carbon there remains enough chromium in solid solution to ensure good resistance to oxidising acids. So for hot 10% nitric acid the high chrome iron has a resistance equal to 18% stainless steel. In fact it has a satisfactory resistance to nitric acid at all strengths and all temperatures except for strong boiling acid when its behaviour becomes irregular.

High chromium cast irons offer little resistance to sulphuric and hydrochloric acids. In this respect they are complementary to the high nickel irons giving good service under conditions where the latter fail and vice versa. Very good resistance is offered by the high chromium irons to sea and tap water attack. Exposures in industrial atmospheres of up to two years have left the surface still bright.

As distinct from the austenitic cast irons the 33% of chrome iron is strongly magnetic and its co-efficient of expansion is similar to that of unalloyed iron and steel at about 11-12 x 10^6 per degree C. Most of these details are from information published by Sheepbridge Engineering Ltd.

4. High Silicon Cast Irons

This material containing silicon at an average 15% is known under various trade names such as :—

Ironal, Tant Iron, Duriron, Durichlor, Anaciron, Corrosion Acidur, Antiacid, Duracid, Eliante and others. It is unmachinable, except by grinding, and very brittle.

Tests which have been made indicate that the silicon content best able to resist different mineral acids should be as follows :—

10% Sulphuric acid	14·8 to 17·3%.
10% Nitric acid	16·0%.
10% Hydrochloric acid ...	18·5%.

Thus in practice, the limits of chemical composition recommended are :—

Carbon	0·2—1·0%.
Silicon	13·0 to 18·0%.
Manganese	0·25 to 2·0%.
Sulphur	0·05% maximum.
Phosphorus	0·20% maximum.

In this range of composition the following approximate physical properties are obtained :—

Density	...	...	...	6·8
Tensile Strength	...	...	100 N/mm²
Compression Strength	...	530 N/mm²
Brinell Hardness	...	...	450 to 500.
Melting point	...	...	1200° C.
Solid contraction	...	...	3%.

At 13% silicon, the carbon solubility is about 1·75% at the point of solidification. With silicon at 18% carbon solubility is reduced to about 0·8%. Greater quantities of carbon can be held in solution in the molten alloy, dependent upon temperature and melting conditions. However, in view of the graphitising action of silicon, all the carbon is present as graphite in the solidified alloy at atmospheric temperature.

Unless care is taken in the selection of raw materials, in melting and in pouring, trouble is likely to arise through the separation of free graphite or "kish" in the mould, together with the presence of flake form graphite in the structure. This further embrittles an already weak material. In this respect it has been found that graphite in the eutectic form is best.

The B.C.I.R.A. has evolved a method for controlling the form of the graphite in high silicon cast irons. It is accomplished by adding from 0·02% to 1·0% of cerium to the molten metal, immediately prior to pouring. On cooling the graphite separates out as finely divided under-cooled and any excess over the eutectic quantity takes on the nodular formation. High silicon cast iron given the cerium treatment shows improved mechanical properties including less brittleness and a marked freedom from porosity and cavities.

For the cerium to be fully effective the base iron must be low in sulphur. An example taken from British and American patents covering the use of cerium in this way was melted in an indirect arc furnace. The charge was :—

        8 kg pig iron.
        15 kg Ferro-Silicon (46% silicon).
        24 kg Mild Steel Scrap.

The pig iron had the chemical analysis T.C. 3·94%; Si 2·78%; Mn 0·57%: S 0·017%; P 0·029%. 25 kg of iron from this charge was tapped into a ladle and 60 g of cerium mischmetall added. Final analysis obtained was:—

Total carbon	...	...	...	1·19%.
Silicon	...	...	...	15·0%.
Manganese	...	...	...	0·59%.
Sulphur	...	...	...	0·015%.
Phosphorus	...	...	...	0·038%.
Cerium	...	...	...	0·118%.

High silicon irons of this type cannot be machined and any finishing has to be done by grinding. Generally, if the castings can be machined with a cutting tool, it is an indication of unsuitability for withstanding severely corrosive conditions.

The high shrinkage of these alloys causes trouble in founding and recent investigations indicate that this property is related, in some degree, to the state of the carbon. It has been suggested that shrinkage increases as more carbon separates as graphite which is the converse of what happens with ordinary grey cast iron alloys. On the whole, it appears that the higher the purity of the iron-silicon alloy the less is the likelihood of cooling troubles.

High silicon cast irons may be melted in the cupola, reverbera-tory, rotary crucible, electric furnace (arc or high frequency). Of these, the cupola must be considered the least suitable in view of keeping the carbon within the desired limits.

The pouring temperature will depend on the size and section of the castings in production. It should be kept as low as possible, the normal range is between 1220° and 1280° C. High casting temperature favours the formation of large columnar crystals. On the other hand, a low temperature produces a fine grain. Liquid shrinkage is not excessive, the real difficulties result from the high solid contraction together with the low strength. Moulds must be loosely rammed and cores made to collapse readily. Any parts of the mould likely to cause resistance to free contraction must be eased after casting. Care must be taken to prevent the formation of fins which encourage cracking. Avoid use of chaplets where possible ; if used they must be of the same composition as the casting.

Correct design is a fundamental requirement. Sharp angles must be avoided and also sudden changes of section. Use generous fillets. Vessels such as boiling pans, receivers and stills should be made with a curved bottom, as large flat surfaces set up contraction difficulties. All joints in pipe work must be formed by means of flanges or spigot and socket since the alloy is too hard to machine or thread.

One of the troubles is to ensure a sufficient freedom from dissolved gases to produce sound castings. Precautions need to be taken at every stage of manufacture to guard against the introduction of hydrogen to the molten metal. It has been established that the presence of hydrogen and carbon monoxide plays a most important part in causing porosity and unsoundness.

As well as resisting acids, the high silicon cast irons can be used against other corroding media and have stood up well also under conditions of erosion and abrasion, such as for example pipes for handling wet sand. The material cannot be recommended for vessels working under internal pressures of more than 3.5 kg/cm^2. It is also unsatisfactory for withstanding rapidly fluctuating temperatures although the danger of cracking is lessened by the provision of suitable facilities for expansion and contraction. For pumps and valves high silicon cast iron is not recommended if the temperature changes are likely to exceed $40°$C. In the case of large tower sections subject to appreciable temperature change, the height must not exceed $1\frac{1}{2}$ times the diameter.

CAST IRONS FOR RESISTING WEAR

Unalloyed Cast Irons

Any grade of cast iron possessing an all white or chilled structure will resist wear and abrasion better than grey cast iron. For the fully white fracture the Brinell hardness will be about 450 to 500. Hardness will vary according to whether the iron is cast in sand or in a chill mould. Also if the cooling rate is not sufficient to retain all the carbon as carbide then the hardness will fall as the structure becomes more mottled and approaches full greyness. Generally, for sand castings to possess an all white fracture under conditions of normal cooling the silicon content will need to be less than 1.0%. When cast into chill moulds the silicon ratio required to produce graphite free structure will depend on the speed of cooling.

Cast iron with an all chilled structure is not machinable in the ordinarily accepted sense of the term. This fact must be borne in mind when considering its suitability for any given purpose. If shaping or boring has to be done then nothing much beyond an all pearlitic matrix cast iron can be employed in which the combined carbon is just below 1.0%. Such cast irons are not wear resistant in the sense of possessing specially high resistance which is being considered under this heading.

High Chrome-Carbon Irons

Some wear resisting alloys develop their full properties only by becoming work hardened. There are many instances involving wear by abrasion in which the scouring action of the abrasive is not sufficient to induce work hardening. In such cases, the high chromium-high-carbon, wear hardening cast irons find useful application. Their range of composition is as follows :—

Total carbon	...	...	...	2·25—2·83%.
Silicon	...	...	...	0·50—1·50%.
Manganese	...	...	...	0·50—1·25%.
Chromium	...	...	...	24·0—30·0%.
Nickel	...	...	...	minimum.
Iron	...	...	...	balance.

They have a Brinell hardness of from 500 to 550 as cast but can be hardened by heat treatment up to 600 Brinell. The heat treatment, which gives greatly improved wear resistance, consists of heating to a temperature of 1100° C. When this temperature is attained, hold it for an hour. Then the castings are allowed to air cool. This heat treatment causes the formation of very unstable austenite, so unstable that it will transform to harder martensite under mild abrasive or rubbing action.

These high-chromium, high-carbon irons can be rendered softer for grinding or simple machining by annealing at 760 to 790° C. for 12 to 24 hours, followed by air cooling. By this treatment the hardness is lowered to between 350 and 450 Brinell.

Heat Treated Low Alloy Cast Irons

Whilst heat treatment of iron castings has limited application and in general is used for simple shapes only, it may offer a solution to some problems of wear. A composition which can be used is :—

Total carbon	...	...	...	3·3%.
Silicon	...	...	...	1·8%.
Nickel	...	...	...	1·5%.
Chromium	...	...	...	0·4%.

After quenching in oil from 850° C. and tempering at 300°/350° C., a hardness of 400 to 450 Brinell is developed.

Ni-Hard

A completely white cast iron which has the property of air hardening by reason of its relatively high nickel and chromium contents. It is a development of the International Nickel Company and can be made in two grades. High carbon imparts maximum hardness whilst the lower carbon grade is used for greater strength properties. Chill castings possess more strength and are harder than sand castings.

Machining is possible only by the use of special tools, otherwise resort to grinding must be made. Inserts of grey cast iron can be "cast in" where it is required to drill and thread.

All Ni-Hard castings benefit from stress-relieving by heating to between 250° and 300° C. No loss of hardness results but strength is improved. Composition and properties as given by the International Nickel Company are:—

Composition per cent.		*High Hardness*	*High Strength*
Total carbon ...	...	3·2—3·6%	2·8—3·2%
Silicon	...	0·4—0·7%	0·4—0·7%
Manganese	...	0·3—1·2%	0·3—1·2%
Sulphur	...	0·15% max.	0·15% max.
Phosphorus	...	0·3% max.	0·3% max.
Nickel	...	3·3—4·8%	3·5—5·0%
Chromium	...	1·2—1·7%	1·2—1·7%
Molybdenum	...	0·0—0·4%	0·0—0·4%
Specific Gravity...	...	7·6—7·8	
Weight lb. per cu. in. ...		0·275—0·28	
Pattern Makers' shrink age in/ft.		$\frac{5}{32}$—$\frac{1}{4}$	

Hardness :—		Sand Cast	Chill Cast	Sand Cast	Chill Cast
Brinell (Tungsten Carbide Ball)		550—650	500—725	500—600	550—650
Vickers D.P.N. (average)	...	590—700	640—800	540—640	590—700
Rockwell C (average)	...	54— 60	57— 64	52— 57	54— 60
Scleroscope (average)	...	72— 82	75— 90	68— 75	72— 80
Tensile tons per sq. in.	...	16— 19	18— 26	18— 22	22— 27

Tellurium Additions

Tellurium introduced to molten cast iron promotes the formation of a hard chill either with or without the aid of metal chillers. It increases the amount of massive carbide in the chill so that a higher hardness and better wearing properties are imparted.

The chilling effect of tellurium does not persist in the re-melted scrap. Thus segregation of scrap, essential with chromium additions, is quite unnecessary. The hardness developed in a tellurium chill is about 450 Brinell.

There is, however, a warning based on evidence put forward by the B.C.I.R.A. that the use of scrap cast iron, containing tellurium, in furnace charges, may cause trouble. It appears that traces of tellurium may persist after remelting in ratios as small as 0·0007%. Such residual quantities of tellurium may give rise to a detrimental arrangement of the graphite which causes very low strength, particularly in grey iron castings of relatively heavy section which cool slowly.

For direct introduction of tellurium into molten cast iron, the best method is to enclose the element in copper tubes or to employ a tellurium copper alloy. Copper tubes containing weighed quantities of tellurium are available. When treating small volumes of iron the required number of TELLURIUM TUBES can be placed in the bottom of a hand shank and the iron tapped on to them.

In view of the relatively low boiling point of tellurium and its great affinity for oxygen, it is advisable not to allow the tubes to float on the surface. If they do the effect may be lost and the fumes evolved will be a nuisance. The most effective and trouble-free method of making the addition, especially in crane ladles, is to wire or wedge the tubes to a stirring rod and plunge them well below the molten surface.

Successful use of tellurium as a chilling medium is dependent on control, especially with respect to :—

1. Grade of cast iron to be treated.
2. Quantity of tellurium used.
3. Effectiveness of the method of making the addition.

The degree of chill depth for any given tellurium treatment would depend on the character of the iron. Thus a low silicon iron will chill much deeper or in much heavier sections than a soft grey iron used for ordinary light engineering castings. Small percentage additions of tellurium only are needed and 0·02% is about the maximum ever required.

Besides direct addition of tellurium to molten cast iron it is possible to impart a chilled surface to iron castings by painting the mould or core, adjacent to the area desired hard, with a dressing containing tellurium. TELLURIT is a mould paint of this kind and is available in both water and spirit based versions.

302

INOCULATED GREY CAST IRON

Cast iron is sensitive to section thickness ; that is to say, the physical and mechanical properties will vary with section or, more correctly, with cooling rate. Arising from this characteristic, it is difficult to predict accurately the mechanical properties of a casting and variations in surface hardness, even hard spots will certainly occur.

The classical means of achieving the desired physical properties and structure of cast iron, in relation to section, has been, and to a large extent still is, to change its chemical composition. For practical reasons, a compromise must be expected because it is impossible to adjust the chemical composition to give optimum properties in each individual casting. Difficulties also arise when a foundry is attempting to make various types of casting of different section thickness from one type of cast iron alloy.

Inoculation

Briefly, inoculation of grey iron is a means of reducing sensitivity to section, thereby enabling the foundryman to produce castings of more consistent quality. While inoculation does not compensate for an unsuitable iron composition, it does greatly reduce the effect of small variations in chemical composition and hence simplifies the function of metal production and control.

The physical properties of grey iron are largely determined by its micro structure. Since variations in micro structure are reflected in physical property variations, the problem of improving physical properties and achieving greater consistency of micro structure must be tackled by some method of improving the micro structure. This is essentially what inoculation does.

Desirable physical properties include good machinability, high tensile strength coupled with toughness, pressure tightness, resistance to wear and useful hardness values. Above all, the effect of casting section on these properties must be minimised, and consequently the micro structure must be as uniform as possible throughout the casting, whatever the variations in section.

Castings produced from iron which has not been inoculated are likely to reveal a white or mottled structure, particularly in thin sections. Such irons frequently show an interdendritic distribution of the graphite flakes coupled with a coarse "eutectic cell" structure, in heavier sections. It is the chill structure which largely accounts for complaints regarding inconsistent machineability.

The reason for these variations lies in the difference in the rate of cooling between the outside of the casting and the interior.

303

This is unavoidable but inoculation does minimise the influence of cooling rate on the micro structure; thereby producing more or less constant properties from centre to edge of a given casting.

What Inoculation Does

In the metallurgical sense, grey cast iron is a brittle material, and even those alloys which come under the category of high duty iron, possessing high tensile strength, may not have increased toughness. In thin sections, the latter possesses hyper-section sensitivity and exhibits under-cooled graphite structures at the surface. The resulting structures are particularly brittle to impact loads, which under service conditions, can off-set any improvement obtained in the bulk of the material.

Inoculation has a marked improvement on toughness, though this property is difficult to define and even more difficult to measure quantitatively.

It is known that a relationship exists between impact strength and the "eutectic cell count" of a grey cast iron. The greater the number of eutectic cells per unit area, the higher is the impact resistance of the cast section. Increasing the number of eutectic cells for a given area is a major outcome of inoculation and for this reason therefore inoculation always increases transverse strength and can effect a marked increase in deflection under load. Since inoculation refines the "grain size" (i.e., increases the eutectic cell count) and also promotes the formation of Type A graphite, the tensile strength is also improved. A very important feature is the reduction in chilling characteristics, thereby greatly reducing surface hardness and producing thin sections free from iron carbide.

While a great deal is known about the theory of inoculation, it is not yet fully understood. Some of the evidence is conflicting regarding the effects of different types of inoculant on various types of iron. The origin of these differences in response to a given inoculant originates in variations of chemical composition, charge composition and melting conditions existing in the cupola. With electric furnace melting, such factors as bath temperature, holding time and general melting routine can all affect the response of the iron to a given inoculant.

A considerable amount of work has already been carried out in order to evaluate these inter-related factors and this has led to the introduction of INOCULIN 10. This product has been shown to give more consistent results over a wider range of iron alloys than any other inoculating material currently available.

Many currently available inoculating agents suffer from practical disadvantages ; calcium silicide is very good for high steel mixes but is not particularly potent in inhibiting chill in thin sections. Its major disadvantage, however, is that it produces a viscous slag on the surface of the ladle which is difficult to remove and all too frequently results in slag inclusions in a casting.

Ferro-silicon is a good inoculant for inhibiting chill but it is not very effective in heavy sections. Quite large additions have to be made which tend to lower the metal temperature, and this also produces a slag on the metal in the ladle. A serious disadvantage of ferro-silicon from the practical point of view is that the inoculation effect appears to be controlled by the amounts of aluminium and/or calcium present as impurities in the ferro-alloy. Indeed, it has been suggested that without the impurities ferro-silicon would be virtually worthless as an inoculant. The amounts of these elements present can vary widely depending on the origin of the ferro-alloy but most varieties used for inoculating purposes will contain between 1·0% and 1·5% of aluminium and up to 0·5% of calcium. Some grades can have up to or even beyond 2·0% of aluminium but in these quantities the possibility of "pin-holing" exists due to promotion of a metal-moisture reaction from retained aluminium in the iron. For consistent inoculating effect therefore it would appear desirable to obtain a consistent ferro-alloy: the inoculating effect due to ferrosilicon is also reported to fade quite rapidly, which can be a marked disadvantage if using large ladles or casting heavy sections.

One of the big advantages of INOCULIN 10 is that it produces a dry dross that is light and very easy to remove from the ladle. This dry dross, moreover, produces a reducing cover which serves to minimise the fading effect associated with other types of inoculating agent.

The Application of INOCULIN 10

The method of adding an inoculating agent, as distinct from the quantity added, is very important. For the best results INOCULIN 10 must be added in a steady stream to the metal as it leaves the cupola spout and the time taken for the total quantity to be added should cover at least 50% of the tapping time. Initial trials can be carried out by hand but for the most consistent results with minimum additions, a mechanical dispenser is highly recommended. The electric vibrator type of dispenser is commonly used.

INOCULIN 10 has revealed that it is up to eight times more efficient than ferro-silicon, and twice as efficient as calcium silicide or other commercially available products. As little as 600 g of INOCULIN 10 per tonne of iron has been equally as effective as 4800 g per tonne of ferro-silicon. In no case has it been necessary to introduce more than 2400 g INOCULIN 10 per tonne of iron in order to effect maximum chill reduction, and in this particular case the charge was produced from a very high steel mix.

The Influence of Base Iron Composition

A major factor which influences the response of an iron to a given inoculant is the composition of the base charge. For example, a soft grey iron (produced from 40% pig iron, remainder returns and bought scrap) will, after treatment with INOCULIN 10, be less likely to chill and form under-cooled graphite structures. As a result, castings will be tougher and have better and more consistent machining characteristics. In castings of up to 25 mm section, inoculation will serve to increase the eutectic cell count, with the result that the main body of the casting will be tougher. Above 25 mm section, the structure begins to coarsen rapidly and physical properties will fall off. Inoculation will not prevent this happening with a charge composition of this type.

By progressively increasing the steel content of the charge the effect of increasing section on coarsening of the structure is reduced, provided that the iron is inoculated. INOCULIN 10 has a distinctly marked effect in promoting this characteristic, while at the same time, the hardness of castings does not noticeably increase. The iron is also less sensitive to chilling in thin sections than is the case with uninoculated iron.

With increasing steel content, the total carbon content is progressively reduced, thus the volume of graphite formed is less and, as a consequence, the iron is potentially stronger and tougher. Again, inoculation of the charge with INOCULIN 10 ensures that this potential is realised.

Thus, whatever the basic composition of the iron charge, inoculation enables the foundryman to produce a more consistent product showing an all round improvement in mechanical properties and in chill reduction.

THE WEDGE CHILL TEST

Importance of Control

In jobbing foundries, where the value of individual castings is often great, and in mechanised foundries, where a considerable value of castings can be produced in a very short time, the use of an unsuitable iron can in either case result in very substantial losses. Some means, therefore, of testing the quality and so suitability of an iron as it comes from the cupola is thus very desirable. A technical committee (T.S.6) of the Institute of British Foundrymen was appointed in 1944 to report on suitable testing methods, and one of their several excellent recommendations was the Wedge Chill Test. Much of the information given here is abstracted from the report of this sub-committee.

Chilling Indications

Different types and compositions of cast iron usually exhibit well-marked differences in chilling tendencies when quickly cooled from the molten state. This chilling tendency is related to melting conditions, chemical composition, particularly as regards carbon and silicon contents, and also to the section thickness cast. For example, in a casting of normal thickness, one iron may exhibit no chilling, another grade might produce a casting having a completely white fracture, while an intermediate grade could produce a casting having a fracture partly grey and partly white. Thus, by casting an iron under carefully controlled conditions of section thickness, etc., a rapid means of assessing its suitability or otherwise for a certain type of casting is provided. The wedge test will also indicate the degree to which depth of chill is reduced after inoculation of the iron.

Wedge Dimensions

In order to minimise the time required for carrying out the test and also the cost involved, the wedge chosen should be the smallest possible consistent with giving a reasonable proportion of white to grey fracture with the particular type of iron in use. There are eight standard sizes of wedge, from A, the smallest, to H, the largest, as detailed in Fig. I and Table I. Table I also shows the time required to complete the test.

An iron required for thin sectioned castings will usually have high total carbon and high silicon contents, in order to avoid chilled sections in the castings themselves. Such irons would, therefore, show little, if any, chill in the test pieces, and in testing, one of the smaller wedges, A, B or C, would be chosen. On the other hand, in testing an iron required for heavier section castings, requiring high strength or high hardness, and which would show a completely white fracture in small test pieces, one of the larger wedges, D, E. F, etc., would be used.

Sand for Mould

To minimise variations through sand conditions, it is recommended that wedges be cast in moulds made of moisture-resistant oil sand. For casting, the moulds can be set in a moulding box, and backed up with loosely-packed moulding sand. Stocks of moulds should always be available.

Testing Procedure

A sample of the metal to be tested is cast into an appropriate size of wedge mould. The pouring temperatures should be standardised, and preferably checked by means of a pyrometer. When the metal has solidified, the wedge should be removed from the mould at a dull red heat—about 600° C. (1100° F.)—and quenched in water. The base side of the test piece should be the first to enter the water, and the piece moved quickly about the water to avoid spurting by the steam. On removing from the water, the wedge is broken up by a breaking block, as shown on page 311.

The fractured section will then clearly indicate a white area at the tip, as indicated. It is the measurement of the base dimensions, W, in millimetres, which is taken as the wedge reading. Page 311 also shows a useful device for making quick measurements.

Applications of the Test

The wedge chill test is of the greatest possible value in controlling the quality of iron to meet required specifications. If, in running a routine test, a variation in the depth of chill appears, it indicates that a change in the composition of the iron has occurred ; provided that cupola operation, pouring temperature and mould conditions have remained constant. The test also indicates the efficiency of an inoculating treatment ; or whether the effect of inoculation has faded.

Apart from its usefulness as a means of checking the consistency of metal as it comes from the cupola, the test may also be used to gauge the suitability of an iron for any specific purpose. The depth of chill obtained can be compared with the depth of chill in wedges cast previously from an iron of a composition known to be suitable for the job in question. For example, if on comparison the depth of chill produced by the iron being tested is greater than that in the control wedges, its composition can be adjusted by making suitable additions of silicon to the ladle.

In the manufacture of high duty types of cast iron with charges containing steel scrap, the wedge test provides an almost indispensable means of control. These types of iron require inoculation treatment at the cupola spout, and the difference in degree of chill shown by wedge tests taken before and after inoculation is a measure of the effectiveness of this treatment.

INOCULIN 10 will reveal its efficiency very well by taking wedge chill test samples before and after inoculation. A guide to the suitability of the inoculated iron, as judged by the wedge test, is that the chill width should not exceed one-third of the average cross sectional thickness of the castings which are to be made. For example, iron which gives a chill measurement of 6 mm in the wedge should not be poured into castings which have a section thickness of less than 18 mm. This ruling applies where maximum tensile strength is required but for general purposes, a wedge value of one-quarter of the average cast section thickness can be adopted.

TABLE 1

Wedge	Base (b)		Height (h)		Length (l)		Time taken to complete test
	mm.	in.	mm.	in.	mm.	in.	
A	6	¼	11·1	$^{7}/_{16}$	57·1	2¼	35 seconds
B	12·7	½	22·2	$^{7}/_{8}$	101	4	40 seconds
C	19	¾	38	1½	127	5	1 minute
D	25·4	1	57·1	2¼	127	5	1½ minutes
E	38	1½	85·7	$3^{3}/_{8}$	152	6	3 minutes
F	51	2	120·6	4¾	152	6	10 minutes
G	63·5	2½	165·1	6½	203	8	15 minutes
H	76·2	3	215·9	8½	203	8	20 minutes

Fig. 1.

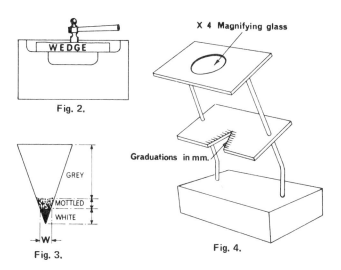

Fig. 2.

GREY

MOTTLED

WHITE

←W→

Fig. 3.

X 4 Magnifying glass

Graduations in mm.

Fig. 4.

RAMMING MIX OR PATCHING FOR CUPOLA LINING

Ganister rock—$\frac{5}{16}+20$ mesh	40%
Fine silica sand	40%
Fireclay of maximum refractoriness and plasticity...	20%
Moisture	7% approximately

When used as a patching or daubing material for cupolas, all the slag from the previous heat must be removed from the brickwork or lining, otherwise when the furnace is in blast the old slag will melt and allow the new patching material to come away.

SECTION V

DIE-CASTING

GRAVITY DIE OR PERMANENT MOULD CASTINGS

General Notes

In this method of casting, molten metal is poured (under "gravity") into a metal ("permanent") mould or die. The metals usually cast are primarily those aluminium alloys that are suitable for this technique, some copper base alloys, e.g. gravity diecasting brass and aluminium bronze, and occasionally zinc, tin, or lead (battery plates) alloys: some ferrous diecasting is also carried out. The die itself can be made of any metal with a similar or higher melting point than the metal being cast, but for a number of reasons a good, close-grained haematite type cast iron is most commonly used. Cores are usually made from heat-resisting steel, but for some castings, e.g. cocks, etc., increasing use is being made of non-re-usable cores made from resin or silicate-bonded sand. This method introduces its own practical difficulties in that the rates of heat extraction from the cast metal as between the metal die and the sand core are vastly different, and unfed shrinkage associated with the sand cored face is common.

Cast Iron Composition suitable for Gravity Dies

A chemical analysis which works well in practice lies within the following limits :

Total carbon	...	3·0—3·3%.
Silicon ...	...	1·2—1·6% according to section thickness.
Manganese	...	0·8—1·2%.
Phosphorus	...	0·3%.
Sulphur ...	...	0·12%.

The heat resisting properties of this material can be improved by making an addition of 1% Simcronite 2 to the molten iron. Dies should be cast with metal at as high a temperature as possible, and it would be an advantage to cast into a heated mould, so as to get more uniform cooling. Before using it is advisable to stress anneal the castings by soaking for at least one hour at 790°C and cooling down slowly in the furnace.

Removable Steel Cores

Cores are constructed from heat resisting steel of which there is a variety of grades. A suggested composition is:

Total carbon ...	...	...	0·3—0·4%.
Silicon ...	...	...	0·3—0·4%.
Manganese ...	...	...	0·2—0·3%.
Chromium ...	...	...	3·0—3·75%.
Vanadium ...	...	...	0·45%.
Tungsten ...	...	...	8·0—9·0%.
Cobalt ...	...	...	1·0—1·5%.

The heat treatment for this steel is: heat slowly to 890°C and quench i.. oil. Temper at 550—580°C by heating slowly and cooling in still air.

Coating the die

It will be appreciated that when molten metal is cast into a relatively cool die the rate of heat extraction, and hence the speed of solidification, can be very fast indeed. So much so in fact that unless special precautions are taken it would be almost impossible to cast any but the most simple shapes. Control of transfer of heat from metal to die is achieved, firstly, by dressing the die with an insulating refractory coating and, secondly, by control of metal and die temperatures. In practice the die-dressing assumes the greatest importance, and temperature control centres mainly around control of pouring temperature. A smoothly cycling die will normally reach a satisfactory temperature balance with its surroundings, and this should be the aim of the operator.

Before casting begins the die is heated to a temperature of about 110—130°C for the application of the refractory dressing by spray gun. If the die is much hotter than this the coating will "frizzle" and leave a rough dressing, while if it is cooler the dressing may "run". Coating a die properly is very much an art. Obviously where maximum insulation is required (runners and risers particularly) the dressing must be applied more thickly, and where maximum chilling is necessary the coating may be only very thinly applied or even subsequently removed entirely. The graduation between maximum chilling and maximum insulation is usually the product of experience both with the die and the dressing being used and the general principles of the process. A "coarse" dressing will give maximum insulation while a "smooth" dressing, while giving a smoother casting finish, will not allow the metal to "run" so easily and may not therefore be so suitable for thin-sectioned castings. Only the die cavity is dressed, and any coating falling on the mating faces or on core prints, etc., is removed. These faces can be treated with a graphite in water dispersion if necessary to improve die operation.

After applying the refractory coating the die is re-heated to about 300—400°C to frit the dressing into position and provide sufficient die superheat to begin casting. Dies should always be above equilibrium temperature at the commencement of a run: if attempts are made to work the temperature up by casting metal into them, the usual results are "stickers" and rapid removal of the die dressing. Small removable cores can rapidly overheat, and it is usual either to use duplicate cores (one set cooling while the other is in the casting) or to dip the entire core into a dilute suspension of graphite in water. A small amount of graphite is picked up each time and will help in providing easier extraction.

Dies for copper base diecastings may not be coated with a refractory dressing as discussed above — a lot will depend on the complexity of the casting. Because of the higher casting temperature, however, die temperature fluctuation will be greater, and it is usually necessary to quench the whole die in an aqueous graphitic dispersion as described for cores. Again, some graphite is picked up each time and forms an essential part of the technique.

In ferrous diecasting the dies may be coated with an insulating— lubricating dressing and, in addition, may be "sooted" with an acetylene flame.

Die Coatings
(a) White insulating dressings:
    DYCOTE 39 (smooth), 34 (coarse), and 140 (medium).
(b) Graphite in water dispersions:
    DYCOTE 11, 38.
(c) Graphite in oil dispersion:
    DYCOTE 40, 400, 410.
(d) Combined insulating/lubricating dressing:
    DYCOTE 36.

Melting Procedure for Aluminium Alloys
Because of the nature of the diecasting process, complex and sophisticated metal techniques are not usually required. Only occasionally is it necessary to obtain complete freedom from dissolved gas in the metal or a high level of grain refinement. The rapid rate of solidification usually ensures that any gas porosity is fine and adequately dispersed and also that grain structures are small and equiaxed. Thus the only treatment accorded the usual bale-out type of furnace is to dross off periodically with COVERAL 11 using about 0·5% of the weight of metal. Indeed, it is frequently advantageous to make deliberate additions of gas to the metal so

that isolated shrinkage unsoundness in the casting is offset by the dispersed gas porosity. To accomplish this, DYCASTAL is added either to the furnace as a plunged tablet or as a "pinch" in the ladle before it is filled.

When, for special quality reasons, it is necessary to provide a full degassing and grain-refining treatment, the technique outlined for sand castings should be used.

The rapid solidification rate in a metal die will also ensure a fair degree of modification of the eutectic in the aluminium—silicon alloys such as LM6, especially in the thinner sections. Where extra modifying potential is required, metallic sodium may be added as NAVAC or NAMETAL or a modifying flux such as COVERAL 36A or 29A may be used.

Special Aluminium Alloys

Under this heading come alloys such as Birmabright (LM5) and "Y" Alloy (L35) for which the technique is as follows:

Melt down the ingot and process scrap under a layer of COVERAL 65, superheat to a maximum temperature of 750°C, and skim. Add a further 2% of the charge weight of COVERAL 65. Allow the flux to fuse on the surface, followed by a thorough puddling of the "wet" slag for 3 to 4 minutes, employing for the purpose a skimmer tool with a rotary motion. Any remaining "islands" of slag or dross should be skimmed off and the sides of the crucible at the metal level gently scraped. Ladling out then proceeds in the usual manner.

In the case of a bale-out furnace working continuously with more ingot metal and scrap being fed in from time to time, the routine outlined would be repeated at intervals dependent upon the rate at which the metal is consumed.

These alloys are susceptible to grain growth, even when cooled relatively quickly as in gravity die-casting. To ensure the production of a fine grain size, NUCLEANT 2 tablets are plunged on a basis of 0·25% by weight.

Although the technique outlined for these special alloys is also suitable for Lo-Ex (LM13), there is an alternative procedure which may be considered simpler. COVERAL 11 is used in the bale-out furnace periodically, purely for cleansing and drossing off. Degassing with DEGASER 190 tablets is carried out at prescribed intervals dependent upon local conditions.

Slush Casting

This is generally carried out in pure aluminium and applied to such components as spouts for kettles and teapots. Common troubles experienced are cracking and sticking to the die.

Pick up of iron is a fruitful cause of cracking so it is best not to use a steel crucible for melting. In addition, all plungers and other steel tools introduced into the molten metal should be well-coated with DYCOTE 36 and dried before use.

Melt down and dross off the pure aluminium with COVERAL 11. NUCLEANT 2 tablets should be plunged at intervals on a basis of 0·25% by weight to counteract excessive grain growth to which pure aluminium is prone. Suitable intervals for applying the NUCLEANT and drossing off will depend on the rate at which the molten metal is used and can coincide with the make up of the bath by additions of fresh ingot or returned scrap. A suitable pouring temperature will be something of the order of 720°C.

Aluminium Bronze

For gravity die-casting in this alloy the 88/9/3, copper—aluminium—iron composition is widely used. Melting is done with the aid of ALBRAL 3 which exerts a powerful solvent action on oxides, including alumina. The latter is formed continuously at the high temperatures involved in the casting of an aluminium bronze alloy, but a protective cover will inhibit its formation. It is advisable to work in small quantities of ALBRAL from time to time, a plunger or similar tool being employed for the purpose. The metal may be further deoxidised by plunging one DEOXIDISING TUBE E3 to each 50 kg of metal.

Under conditions of working, fresh metal has to be added as casting proceeds. Thus, the treatment with ALBRAL and DE-OXIDISING TUBES must be repeated at suitable intervals. The exact nature of the technique employed will depend on whether a bale-out furnace is used to hold a reservoir of metal which is replenished from a separate melting furnace.

An average pouring temperature is around 1200°C, which makes it necessary to use special hand ladles of heat-resisting alloy steel. For the same reason the dies are sometimes sunk in special steel although close-grained cast iron is employed satisfactorily.

For coating the gravity die blocks DYCOTE 36 is suggested. The steel cores, especially if of thin section, need cooling down in between each cast. This is conveniently done by dipping them into an aqueous suspension of DYCOTE 11. The graphite coating thus applied imparts the desired lubricating properties, and assists in the extraction of the cores.

Turbulence must be guarded against when pouring the die, bearing in mind that aluminium-bronze is a dross forming alloy. The relatively high solidification shrinkage must also be taken into consideration in allocating the dimensions of feeding risers. In the effort to avoid turbulent flow and agitation of the molten alloy in the die cavity, full advantage should be taken of dies which lend themselves to tilting during pouring, the die being brought back to the vertical position as metal enters the feeder.

A precaution against cracking is to rotate cores and pins whilst the metal is being poured. These should be withdrawn as soon after casting as possible. In fact, the whole die assembly should be released as quickly as is practicable after the component has set.

DEOXIDISING TUBES L have been applied successfully to the gravity die-casting of aluminium-bronze, particularly in cases where cracking has been experienced due to the presence of impurities, especially bismuth. Aluminium—bronze is very hot short at its critical temperature, and the presence of impurities tends to interfere with the normal mechanism of solidification, leading to minor or major cracking. DEOXIDISING TUBES L also influence grain refinement and act as a degasser. A DEOXIDISING TUBE LI containing 5·5 g lithium, is normally sufficient to treat 75 to 100 kg charge. The tubes must be completely immersed.

Gravity Die-casting of Brass

The chemical specification for die-casting brass is substantially 60/40 copper—zinc alloy, containing also up to 0·5% aluminium and from 0·5—1·0% of lead.

Where multiple piece dies are used they are taken apart and dipped in a water suspension of DYCOTE 11 after each pour. In this way the dies are kept cool and the tendency for build-up to occur on the face of the dies is considerably reduced. The steel cores are also dipped in the suspension of DYCOTE 11 after each cast. An alternative coating for the die blocks is DYCOTE 36, or DYCOTE 61 which is a graphite-based dressing developed to reduce zinc oxide build-up on the die walls.

Melt the brass under a protective flux cover of 1% by weight of ALBRAL 3. Thereafter about 0·5% of the charge weight is stirred into the melt at intervals, preferably at the time of adding fresh alloy to the holding furnace. The dross which forms on the surface of the melt can be skimmed off following each application, or pushed aside while ladling out. Superheat the bath to a temperature of about 1100°C and pour the dies with metal at about 1050°C.

319

Gravity Die-casting of Gunmetal

Gunmetal alloys have not been employed very extensively for gravity die-casting in Great Britain. There are a number of reasons which include short die life, extraction difficulties, and sub-surface porosity, etc. Even so, satisfactory gunmetal die-castings can be produced provided the component is small and of simple design.

In melting down the initial bale-out charge, CUPREX 1 blocks should be employed on a basis of 1% of the charge weight. Thereafter melting will be continuous with fresh ingot metal and returned scrap being added from time to time to maintain the bath of molten alloy. Subsequently REGENEX 6 may be plunged at appropriate intervals to keep the liquid metal clean and degassed.

Following the degassing operation, strong deoxidation is carried out by plunging one DEOXIDISING TUBE DS4 to each 25 kg of melt. This will also have the effect of promoting fluidity to a maximum degree.

Pouring temperatures will range from $1150°$ to $1200°C$, and the die should be heat-resisting cast iron or steel to stand up successfully.

DYCOTE 36 is recommended for spraying the die blocks, which should be sunk with as much taper as dimensional tolerance will allow, otherwise extraction difficulties will become a major problem.

Eliminate any tendency to build up back pressure by cutting venting grooves in the die at as many points as design permits.

Before casting commences always pre-heat the dies to a working temperature of $350-400°C$ with the idea of limiting thermal shock and blowing.

Gravity Die-casting of Lead and Lead Base Alloys

Melt down under a cover of 1% of the charge weight of PLUMBREX 2 which is also used for drossing off purposes prior to pouring. PLUMBRIT 1 tablets should be plunged occasionally for thorough cleansing (one tablet per 100 kg).

The die face may be coated with an insulating dressing as used for diecasting of aluminium alloys (see appropriate section) if feeding problems are anticipated and directional solidification is required. For simple castings, however, it is often sufficient simply to apply a coating of silicone (most conveniently from an aerosol) to the bare die surface.

Battery plate manufacturers typically employ a highly insulating coating because of the very thin grids they are making, and this is usually based on cork flour with a silicate binder. Individual casters frequently add additional materials, cellulose binder derivatives being most common. Battery grid casting is a very specialised section of gravity diecasting of lead, and most producers have evolved their own individual techniques.

Gravity Die-casting of Magnesium Base Alloys

It is best not to attempt the production of complicated components containing many changes of section. Dies normally used for aluminium alloys need to have the runner-riser system enlarged. Magnesium alloys are more prone to shrinkage cavitation.

The following observations apply to the bale-out furnaces fitted with steel crucibles but where no protection is afforded the molten alloy by hooding and a continuous atmosphere of sulphur dioxide. Under such conditions it is essential that the surface of the melt be covered by a thin "barrier" of INERTEX, thus reducing burning to negligible levels.

Into the pre-heated crucible place MAGREX 60 on a basis of about 3% by weight of the final charge. Add pre-heated magnesium ingot and/or scrap, in stages, until the melt is complete. Keep the flux cover intact and make good any exposed areas by a further addition of MAGREX 60. Keep the melt to a working temperature of $730-750°C$ for the refining stage.

For refining, the MAGREX 60 fluid slag is thoroughly stirred into the bulk of the metal over a period of three to four minutes. If properly conducted the surface of the magnesium assumes a bright lustre. A "frothy" appearance indicates the need of continued refining with a fresh introduction of fluxing compound.

The slag loaded with entrained oxides and impurities gradually sinks although some will cling to the walls of the pot. Atmospheric attack upon the surface of the melt should be curtailed by adding a little fresh flux.

Grain refinement is carried out by plunging NUCLEANT 200 tablets at a rate of approximately 125 g per 50 kg of magnesium alloy. In most alloys these tablets have a sufficiently high density to sink of their own accord. Simply displace the flux on the surface of the melt and place the tablets on the exposed surface. Allow the reaction to take place steadily over a period of two to three minutes, after which the charge is allowed to settle for ten minutes under its protecting cover of flux compound.

Finally, skim off the surface flux in stages. As the metallic surface is exposed, cover the area immediately with INERTEX, so preventing any tendency to burn. Ladling out can then commence. The die cavity is flushed with sulphur dioxide, delivered under pressure from a cylinder, prior to each pour. The die blocks are dressed with a strongly adhering coating such as DYCOTE 140.

Gravity Die-casting of Tin

As a covering flux during melting use STANNEX 101 which can also be applied at intervals for cleaning up the melt when fresh ingot or scrap is added. This should be placed on the surface of the molten tin and plunged slowly as directed.

The temperature of the molten tin bath should not rise above 280°C, and a temperature of between 260° and 270°C is suitable for pouring into the die. DYCOTE 39 is used for coating the die or silicone, as described for lead alloys, may be used.

PRESSURE DIE-CASTING

General Notes

The following points are of importance for the production of sound pressure die-castings in whatever metal or alloy.

1. Use as low an injection temperature as will ensure complete filling of the die cavity.

2. Maintain a suitable average die temperature free from sharp local temperature gradients.

3. Position the gate so that the injected metal does not impinge directly upon die or core surfaces.

4. Employ an injection pressure high enough for a strong pressure pulse to be transmitted through the solidifying casting at the end of the injection stroke. Where these conditions are observed the incidence of porosity, shrinkage voids and surface imperfections is reduced to negligible proportions.

Composition of Die Steels

It is usual to employ steels of heat-resisting composition for the construction of pressure dies and cores where required.

Die-casting die steels are usually machined in the soft annealed condition and heat treated after machining. The temperature and length of time required for the various steps in heat treatment depend on the type of steel and the size and intricacy of the die being treated. It is best to obtain the advice of the makers of the steel and their instructions for suitable heat treatments.

Some typical chemical analyses of tool steel for die-blocks are given in the following table taken from the book on die-casting by Charles O. Herb.

Table I

COMPOSITION OF STEELS COMMONLY USED FOR DIE-CASTING DIES

Chemical Element	Letters Designating Different Types of Steel							
	A	B	C	D	E	F	G	H
Carbon ...	0·20–0·50	0·40–0·50	0·40–0·50	0·30–0·40	0·30–0·40	1·40–1·50	0·35–0·45	0·25–0·35
Manganese	—	0·65–0·80	0·40–0·80	0·20–0·35	0·20–0·35	—	0·20–0·35	0·20–0·35
Silicon ...	—	0·30	0·30	0·8–1·0	0·8–1·0	—	0·35	0·35
Chromium	—	0·75–0·90	2–2·50	4·75–5·75	4·5–5	12–13	2·5–3·5	2·5–3·25
Vanadium	—	—	0·15–0·30	—	—	—	0·30–0·60	0·30–0·60
Nickel ...	—	1–1·50	—	—	—	—	—	—
Tungsten ...	—	—	—	4–5	0·75–1·25	—	8–10	12–16
Molybdenum	—	—	—	—	1–1·50	0·50–0·60	—	—
Cobalt ...	—	—	—	0·50	—	3–4	—	—

Table 2

BRINELL HARDNESS OF DIE STEELS LISTED IN TABLE I

Base Alloys for which the Listed Die Steels are Used	Letters Designating Different Types of Steel							
	A	B	C	D	E	F	G	H
Tin ...	220	170–220	—	—	—	—	—	—
Lead ...	220	170–220	—	—	—	—	—	—
Zinc ...	170–250	170–300	300–402	—	360–402	—	—	—
Magnesium	—	—	387–430	402–444	402–444	—	—	—
Aluminium	—	—	402–444	420–460	420–460	430–517	444–517	444–517
Copper ...	—	—	—	—	—	430–517	444–517	444–517

Metal Treatment for Aluminium Base Alloys

The popular British aluminium base alloys for pressure die-casting include: LM1, LM2, LM4, LM6.

The melting and fluxing techniques for these alloys have already been dealt with (see pages 122 to 132) and in the main these instructions apply here. In some cases modifications are advisable and these are included in the notes which follow.

Aluminium Alloy LM6

As melting down and drossing off flux the low melting point COVERAL 36A is recommended. This can be used for the initial melting down in the bale-out furnace and applied for cleaning up the surface of the charge at appropriate intervals and when fresh ingot metal or scrap are added.

Aluminium Alloy LM2

The instructions given for LM6 apply equally in this case also. Sometimes trouble is experienced due to the casting welding, or sticking, to the die. So far as aluminium base alloys are concerned an opinion has been expressed that up to about 0·8% of iron in the alloy helps to prevent this taking place. Suitable die dressings are, of course, a factor in overcoming this trouble and some experimentation is often necessary.

Aluminium Alloy LM4

The melting and fluxing of this alloy consist in the plunging of DEGASER 190 at pre-determined intervals prior to drossing-off with COVERAL 11. The freezing range of this alloy is between 525 and 625°C. A suitable temperature range for injection is 650–670°C.

Aluminium Alloy LM1

Instructions as for LM4 apply.

Magnesium Base Alloys

If the magnesium alloy is in the form of scrap and is melted down and refined prior to treatment in the bale-out furnace, it should be melted and refined, using MAGREX 60. The technique to be followed here has already been given on pages 158–169.

As the slag is comparatively heavy, much of it sinks to the bottom of the pot, leaving behind a thin but tenacious protective film. About 1% of the charge weight of flux is used. It is customary to apply the flux at intervals which are dictated, mainly, by the rate at which the metal is replenished. From time to time it is necessary to dredge out the spent sludge from the bottom of the crucible.

One big advantage in the case of pressure die-casting magnesium alloys is the fact that the metal is worked at a relatively low temperature, thus curtailing oxidation. The temperature of the magnesium alloy in the bale-out furnace prior to transference to the cold-chamber die-casting machine should be about 680°C. The steel crucible of the bale-out furnace for magnesium base alloys should be of pressed or cast construction. There is no need to employ inhibitors in the form of SO_2 gas for flushing the dies as applies to the gravity die-casting of magnesium base alloys.

As some form of "self healing" surface flux is called for during ladling out, INERTEX is very effective and a light dusting at regular intervals is sufficient to maintain the protective skin over the metal.

Zinc Pressure Die-Casting

The zinc base casting alloy used in the majority of cases is Mazak No. 3 or No. 5 to British Standard Specification 1004, alloys A and B. These contain about 4% of aluminium and some copper and very carefully controlled quantities of magnesium, lead, iron, cadmium, and tin, the balance being zinc.

A recommended procedure for the melting of zinc base alloys such as these is the use of ZINCREX flux in a ratio of about 0·5% of the charge weight. The method of application will depend to some extent on the type of pressure die-casting machine in use. Some machines have a self-contained reservoir and work upon the goose-neck principle. In this the capacity is limited and it is not convenient to apply a fluxing compound to the metal in the reservoir. On the other hand, where the die-casting machines are fed from a separate bale-out furnace ZINCREX can be used regularly at intervals with distinct advantage. Suitable intervals would be when making up the bath with fresh ingot metal or returned scrap. A little flux worked into the surface of the metal at these stated intervals will keep the metal bath clean and free from dross and suspended non-metallics. After working in the flux as described the surface dross is removed.

Injection temperatures should be kept as low as possible consistent with soundness and good surface appearance. A suggested range would be from $400-425^{\circ}$C for Mazak alloys, and the temperature of the dies should be kept somewhere in the limits of $180-220^{\circ}$C.

Pressure Die-Casting of Brass

The alloy most used is a 60/40 copper—zinc brass containing up to 0·2% of aluminium. This alloy melts at a temperature of about 900°C, and die-casting machines for brass are designed on the principle of casting the metal while it is in a semi-liquid or plastic condition rather than in the completely molten state. In this way, the life of the dies is increased and, in addition, cold water is constantly circulated through plates behind the die face to prevent them from becoming heated beyond 350°C. The alloy is injected into the die at a temperature of about 950°C.

If the alloy containing aluminium is employed it is melted under a cover of ALBRAL 3 flux and treated with DEOXIDISING TUBES DS to give additional fluidity. Use 1% of ALBRAL 3 during the initial melt-down period and then stir in 0·5% at regular intervals such as when the bath is replenished by fresh ingot metal or scrap. Two DEOXIDISING TUBES DS1 can be added per 50 kg of melt and pro rata according to the rate of replenishment of the bath.

Small machines which will cast parts with an area up to 45 cm² and a weight up to about 300 g apply a pressure of about 350 kg/cm² on the metal being cast. Medium-sized machines will deal with castings up to 90 cm² in area, and 600 g in weight. Large machines develop a pressure of 700 kg/cm² or even higher.

The largest die-casting machine can be operated at the average rate of 100 shots per hour, while the other machines can be operated at from 1700 to 1800 shots per 8 hour day.

Pressure Die-Casting of Lead

Lead base alloys are among the easiest to pressure die-cast, and their low melting point allows of long die life. Usually the lead is alloyed with antimony to increase hardness. The percentage of antimony added is normally between 10% and 12%. Sometimes 4—5% of tin is also added to improve the fluidity.

For melting the alloy we recommend the use of PLUMBREX 2 as a flux used in a ratio of from 0·5% to 1% of the charge weight.

Pressure Die-Casting of Tin and Tin Base Alloys

Tin, and alloys based on tin, are also easy to pressure die-cast, their low melting point causing little deterioration of the dies.

When commencing a melt, the initial charge should be melted down under a protective cover of STANNEX 101, using sufficient of it to cover the surface area of the melt. Thereafter the application of the flux will depend upon operating conditions and the type of pressure die-casting machine in use. Some machines have a self-contained reservoir and work upon the goose-neck principle. In this the capacity is limited and it is not convenient to apply a fluxing compound to the metal in the reservoir. On the other hand, where the die-casting machines are fed from a separate bale-out furnace, STANNEX 101 can be used regularly at intervals with distinct benefit. Suitable intervals would be when making up the bath with fresh ingot metal or returned scrap. At these times, place STANNEX 101 to the extent of about 0·25% of the charge weight on the surface of the molten metal and press it slowly to the bottom with the aid of a perforated plunging tool. Hold the plunger down until the reaction ceases, stir well, withdraw the plunger, and skim off any dross or residue.

Dressings for Pressure Diecasting

All the metals mentioned in the previous section can, with the exception only of the copper-base alloys, be most conveniently cast using the new water-dispersible lubricants. In average practice about 1 part of lubricant is dispersed in around 40 parts by volume of water to form a creamy emulsion, stable under most conditions. Fairly generous quantities of this dispersion are sprayed on to the die face compared with the quantity of oil-based lubricant previously used. There is, therefore, a marked cooling effect from this, and it may be possible to reduce the flow of cooling water through a water-cooled die or, alternatively, increase the cycling rate of a non-water-cooled die. In addition to the technical and practical advantages of these dressings they are, compared with the oily types, very much cleaner in use since little oil smoke is given off and there is no fire hazard. They are easily removed by conventional degreasing processes and may therefore be used satisfactorily for castings subsequently painted or plated.

For a particularly difficult casting inclined to "stick" in one or more spots or if a local hot spot exists in an otherwise satisfactory die, the effect of these water dispersible dressings can be reinforced by touch swabbing of the critical area(s) with a filled lubricant such as DYCOTE 670 or DYCOTE 40. The fillers are finely divided aluminium flake and colloidal graphite respectively. These filled dressings can, if required, be diluted with paraffin or white spirit and used conventionally as die lubricants using similar swabbing or spraying techniques. Graphite containing dressings are also frequently used on the back of the moving half of the die to provide lubrication for cores, ejector pins, etc.

The pressure diecasting of brass and other copper base alloys normally means, because of the higher temperatures involved, that graphite in some form is essential on the face of the die. DYCOTE 40 is therefore used either as supplied or diluted with a little light mineral oil. It is best applied by fine spray to ensure an even thin coating on the die face. If a water-dispersible graphite containing coating is required, DYCOTE 11 or 38 should be used, a suitable dilution being found by trial and error.

SECTION VI

STEEL CASTINGS

CONTENTS

American Specifications for Steel Castings.

British Specifications for Steel Castings.

Application of Feeding Aids to Steel Castings.

Moulding and Core Sand Mixtures.

Pages 331 to 349 are reprinted from the "Summary of Steel Castings Specifications" by permission of the Steel Founders Society of America.

Pages 350 to 353 are reprinted from British Standards Specification 3100 by permission of the British Standards Institution, 2 Park Street, London W.I, from whom copies of the complete standard may be purchased.

AMERICAN SPECIFICATIONS
FOR STEEL CASTINGS

Carbon and Low-Alloy Cast Steels

For details of the following specifications see pages 333 to 349.

ASTM   A 27-62 Mild to Medium-Strength Carbon-Steel Castings for General Application.

ASTM   A 148-60 High-Strength Steel Castings for Structural Purposes.

ASTM   A 216-63T Carbon Steel Castings Suitable for Fusion Welding for High Temperature Service.

ASTM   A 217-60T Alloy Steel Castings for Pressure Containing Parts Suitable for High Temperature Service.

ASTM   A 352-60T Ferritic Steel Castings for Pressure Containing Parts Suitable for Low Temperature Service.

ASTM   A 356-60T Heavy-Walled Carbon and Low-Alloy Steel Castings for Steam Turbines.

ASTM   389-60T Alloy Steel Castings Specially Heat Treated for Pressure Containing Parts Suitable for High Temperature Service.

ASTM   A 486-63T Steel Castings for Highway Bridges.

ASTM   A 487-63T Low-Alloy Steel Castings Suitable for Pressure Service.

SAE   1962 Automotive Steel Castings.

AAR   M 201-62 Steel Castings.

ABS   Am. Bur. Shipping Steel Castings—1964 Rules Edition—Machinery and Hull Castings.

Lloyds' Register of Shipping 1963—Steel Castings.

High Alloy Cast Steels

For details of the following specifications see pages 343 to 349.

ASTM A128–60 Austenitic Manganese Steel Castings.

ASTM A296–63T Corrosion-Resistant Iron-Chromium and Iron-Chromium-Nickel Alloy Castings for General Application.

ASTM A297–63 Heat Resistant Iron-Chromium and Iron-Chromium-Nickel Alloy Castings for General Application.

ASTM A351–63T Ferritic and Austenitic Steel Castings for High Temp. Service.

ASTM A447–50 Chromium-Nickel-Iron Alloy Castings (25-12 Class) for High Temp. Service.

ASTM 448–50 Nickel-Chromium-Iron Alloy Castings (35-15 Class) for High Temp. Service.

MILITARY MIL–S–16993A December 1954 Steel Castings (12 per cent Chromium).

      MIL–S–867A December 1951 Steel Castings Corrosion Resisting Austenitic.

Pages 333 to 349 are reprinted from the "Summary of Steel Castings Specifications" by permission of The Steel Founders Society of America.

Explanation of symbols used see pages 341, 342 and 349.

American Specifications for Steel Castings—continued

CARBON AND LOW ALLOY CAST STEELS

Physical Properties

	SPECIFICATION & HEAT TREATMENT			MECHANICAL PROPERTIES—MINIMUM				
Specification	Class	Heat Treatment ‡		Tensile Strength p.s.i.	Yield Point p.s.i.	Elong. in 2" %	Red. of Area %	Other Tests: Bend, Impact Hardness†
ASTM A 27-62	N-1	—		—	—	—	—	—
	N-2	A or N or NT or QT		—	—	—	—	—
	U-60-30	A or N or NT or QT		60,000	30,000	22	30	—
	60-30	A or N or NT or QT		60,000	30,000	24	35	—
	65-35	A or N or NT or QT		65,000	35,000	24	35	—
	70-36	A or N or NT or QT		70,000	36,000	22	30	—
	70-40	A or N or NT or QT		70,000	40,000	22	30	—
ASTM A 148-60	80-40	A or N or NT or QT		80,000	40,000	18	30	—
	80-50	A or N or NT or QT		80,000	50,000	22	35	—
	90-60	A or N or NT or QT		90,000	60,000	20	40	—
	105-85	A or N or NT or QT		105,000	85,000	17	35	—
	120-95	A or N or NT or QT		120,000	95,000	14	30	—
	150-125	A or N or NT or QT		150,000	125,000	9	22	—
	175-145	A or N or NT or QT		175,000	145,000	6	12	—
ASTM A 216-63T	WCA	A or N or NT		60,000	30,000	24	35	Bend—Degrees [90]
	WCB	A or N or NT		70,000	36,000	22	35	[90]
ASTM A 217-60T	WC1	A or NT		65,000	35,000	24	35	[90]
	WC4	A or NT		70,000	40,000	20	35	[90]
	WC5	A or NT		70,000	40,000	20	35	[90]
	WC6	A or NT		70,000	40,000	20	35	[90]
	WC9	A or NT		70,000	40,000	20	35	[90]
	C5	A or NT		90,000	60,000	18	35	[90]
	C12	A or NT		90,000	60,000	18	35	[90]

American Specifications for Steel Castings—continued

CARBON AND LOW ALLOY CAST STEELS—continued

Composition

Specification	CHEMICAL COMPOSITION %—MAXIMUM								
	C	Mn	P	S	Si	Ni	Cr	Mo	Other Elements
ASTM A 27–62	.25(2)	.75(2)	.05	.06	.80	.50	.40	.20	Cu .30
	.35(2)	.60(2)	.05	.06	.80	.50	.40	.20	Cu .30
	.25(2)	.75(2)	.05	.06	.80	.50	.40	.20	Cu .30
	.30(2)	.60(2)	.05	.06	.80	.50	.40	.20	Cu .30
	.30(2)	.70(2)	.05	.06	.80	.50	.40	.20	Cu .30
	.35(2)	.70(2)	.05	.06	.80	.50	.40	.20	Cu .30
	.25(2)	1.20(2)	.05	.06	.80	.50	.40	.20	Cu .30
ASTM A 148–60	—(1)	—	.05	.06	—	—	—	—	—
	—(1)	—	.05	.06	—	—	—	—	—
	—(1)	—	.05	.06	—	—	—	—	—
	—(1)	—	.05	.06	—	—	—	—	—
	—(1)	—	.05	.06	—	—	—	—	—
	—(1)	—	.05	.06	—	—	—	—	—
	—(1)	—	.05	.06	—	—	—	—	—
ASTM A 216–63T	.25(11)	.70(11)	.05	.06	.60	.50(5)	.40(5)	.25(5)	Cu .50(5)
	.30	1.00	.05	.06	.60	.50(5)	.40(5)	.25(5)	Cu .50(5)
ASTM A 217–60T	.25	.50–.80	.05	.06	.60	.50(6)	.35(4)	.45–.65	Cu .50(4)
	.20	.50–.80	.05	.06	.60	.70–1.10	.50–.80	.45–.65	Cu .50(4)
	.20	.40–.70	.05	.06	.60	.60–1.00	.50–.90	.90–1.20	Cu .50(4)
	.20	.50–.80	.05	.06	.60	.50(6)	1.00–1.50	.45–1.50	Cu .50(4)
	.18	.40–.70	.05	.06	.60	.50(6)	1.00–2.75	.45–1.20	Cu .50(4)
	.20	.40–.70	.05	.06	.75	.50(6)	4.00–6.50	.45–.65	Cu .50(4)
	.20	.35–.65	.05	.06	1.00	.50(6)	8.00–10.00	.90–1.20	Cu .50(4)

Physical Properties

Specification	Class	Heat Treatment #	Tensile Strength p.s.i.	Yield Point p.s.i.	Elong. in 2" %	Red. of Area %	Other Tests: Bend, Impact Hardness[7] Impact-Ft.-Lbs.[7]
ASTM A 352-60T	LCB	Z or NT (or QT)	65,000	35,000	24	35	15 @ −50°
	LC1	Z or NT (or QT)	65,000	35,000	24	35	15 @ −75°
	LC2	Z or NT (or QT)	65,000	40,000	24	35	15 @ −100°
	LC3	Z or NT (or QT)	65,000	40,000	24	35	15 @ −150°
ASTM A 356-60T [12]	1	NT	70,000	36,000	20	35	—
	2	NT	65,000	35,000	22	35	—
	3	NT	80,000	50,000	18	35	—
	4	NT	90,000	60,000	16	35	—
	5	NT	70,000	40,000	22	35	—
	6	NT	70,000	45,000	22	35	—
	7	NT	70,000	40,000	22	35	—
	8	NT	80,000	50,000	18	35	—
	9	NT	95,000	60,000	15	35	—
	10	NT	85,000	55,000	20	35	—
ASTM 389-60T	C23	1850 N, 1250 T (Min) 1 hr/in.	70,000	40,000	18	35	—
	C24	1850 N, 1250 T (Min) 12 hr.	80,000	50,000	15	35	—
ASTM A 486-63T	70	N or NT or QT	70,000	36,000	22	30	25[14]
	90	N or NT or QT	90,000	60,000	20	40	25[14]
	120	QT	120,000	95,000	14	30	30[14]

MECHANICAL PROPERTIES—MINIMUM

American Specifications for Steel Castings—continued
CARBON AND LOW ALLOY CAST STEELS—continued
Composition

Specification	C	Mn	P	S	Si	Ni	Cr	Mo	Other Elements
				CHEMICAL COMPOSITION %—MAXIMUM					
ASTM A 352–60T	·30	1·00	·05	·06	·60	—	—	—	—
	·25	·50—80	·05	·06	·60	—	—	·45—65	—
	·25	·50—80	·05	·05	·60	2·00—3·00	—	—	—
	·15	·50—80	·05	·05	·60	3·00—4·00	—	—	—
ASTM A 356–60T (12)	·35(2)	·70	·05	·05	·60	—	—	—	V (6)
	·25(2)	·70(2)	·05	·05	·60	—	—	—	V (9)
	·25	·50—80	·05	·05	·60	—	—	·40—·60	
	·20	·50—80	·05	·05	·60	—	—	·90—1·20	
	·25(2)	·70(2)	·05	·05	·60	—	·40—·70	·40—·60	
	·20	·50—80	·05	·05	·60	—	1·00—1·50	·40—·60	V ·15—·25
	·20	·50—80	·05	·05	·60	—	1·00—1·50	·90—1·20	V ·15—·25
	·20	·50—80	·05	·05	·60	—	1·00—1·50	·40—·60	V ·15—·25
	·20	·50—80	·05	·05	·60	—	1·00—1·50	·90—1·20	V ·15—·25
	·20	·50—80	·05	·05	·60	—	2·00—2·75	·90—1·20	
ASTM A 389–60T	·20	·30—80	·05	·06	·60	—	1·00—1·50	·45—65	V ·15—·25
	·20	·30—80	·05	·06	·60	—	·80—1·20	·90—1·20	V ·15—·25
ASTM A 486–63T	·35	·90	·05	·06	·80	—	—	—	—
	·35	(1)	·05	·06	(1)	(1)	(1)	(1)	(1)
	·35	(1)	·05	·06	(1)	(1)	(1)	(1)	(1)

Physical Properties

SPECIFICATION & HEAT TREATMENT			MECHANICAL PROPERTIES—MINIMUM				
Specification	Class	Heat Treatment #	Tensile Strength p.s.i.	Yield Point p.s.i.	Elong. in 2" %	Red. of Area %	Other Tests: Bend, Impact Hardness†
ASTM A 487–63T	1–N	NT	85,000	55,000	22	40	
	2–N	NT	85,000	53,000	22	35	
	3–N	NT	90,000	60,000	20	35	
	4–N	NT	90,000	60,000	20	40	
	5–N	NT	105,000	70,000	18	35	
	6–N	NT	115,000	80,000	18	30	
	7–N[13]	NT	110,000	70,000	18	35	
	8–N	NT	85,000	55,000	20	35	
	9–N	NT	90,000	60,000	20	35	
	10–N	NT	100,000	70,000	18	35	
	1–Q	QT	90,000	65,000	22	45	
	2–Q	QT	90,000	65,000	22	40	
	3–Q	QOT	105,000	85,000	17	30	
	4–Q	QOT	105,000	85,000	17	35	
	5–Q	QOT	105,000	85,000	15	30	
	6–Q	QOT	120,000	95,000	12	25	
	7–Q[13]	QT	125,000	100,000	15	30	
	8–Q	QT	105,000	85,000	17	30	
	9–Q	QT	105,000	85,000	16	35	
	10–Q	QT	125,000	100,000	15	35	

CARBON AND LOW ALLOY CAST STEELS—continued

Composition

Specification	CHEMICAL COMPOSITION %—MAXIMUM								
	C	Mn	P	S	Si	Ni	Cr	Mo	Other Elements
ASTM A 487–63T	·30	1·00	·05	·06	·80	·50(3)	·35	—	V ·07–·15, Mo+W ·25 Cu ·50
	·30	1·00–1·40	·05	·06	·80	·50(16)	·35	·10–·30	
	·35	1·35–1·75	·05	·06	·80	·50(16)	·35	·25–·55	
	·35	1·00	·05	·06	·80	·40–·80	·40–·80	·15–·30(15)	
	·30	1·00–1·40	·05	·06	·80	·40–·80	·40–·80	·15–·25(15)	
	·38	1·30–1·70	·05	·06	·80	·40–·80	·40–·80	·30–·40(15)	
	·20	0·60–1·00	·04	·05	·80	·70–1·00	·40–·80	·40–·60(15)	V ·03–·10 + B ·002– ·006 + Cu ·15–·50
	·20	0·50–0·90	·05	·06	·80	—	2·00–2·75	·90–1·10(15)	
	·33	0·60–1·00	·05	·06	·80	·50(16)	·75–1·10	·15–·30	
	·33	0·60–1·00	·05	·06	·80	1·40–2·00	·55–·90	·20–·40(15)	
	·30	1·00	·05	·06	·80	·50(3)	·35	—	V ·07–·15, Mo+W ·25 Cu ·50
	·30	1·00–1·40	·05	·06	·80	·50(16)	·35	·10–·30	
	·35	1·35–1·75	·05	·06	·80	·50(16)	·35	·25–·55	
	·30	1·00	·05	·06	·80	·40–·80	·40–·80	·15–·30(15)	
	·30	1·00–1·40	·05	·06	·80	·40–·80	·40–·80	·15–·25(15)	
	·30	1·30–1·70	·05	·06	·80	·40–·80	·40–·80	·30–·40(15)	
	·20	0·60–1·00	·04	·05	·80	·70–1·00	·40–·80	·40–·60(15)	V 03–·10 + B ·002– ·006 + Cu ·15–·50
	·20	0·50–0·90	·05	·06	·80	—	2·00–2·75	·90–1·10(15)	
	·33	0·60–1·00	·05	·06	·80	·50(16)	·75–1·10	·15–·30	
	·33	0·60–1·00	·05	·06	·80	1·40–2·00	·55–·90	·20–·40(15)	

Physical Properties

SPECIFICATION & HEAT TREATMENT			MECHANICAL PROPERTIES—MINIMUM				
Specification	Class	Heat Treatment #	Tensile Strength p.s.i.	Yield Point p.s.i.	Elong. in 2" %	Red. of Area %	Other Tests: Bend, Impact Hardness†
SAE Automotive	0022	A or N or NT	60,000	—	22	—	BHN 187 Max.
	0025	A or N or NT	65,000	30,000	24	30	187 Max.
	0030	A or N or NT or QT	85,000	35,000	16	35	131–187
	0050A	N or NT	100,000	45,000	10	24	170–229
	0050B	QT	80,000	70,000	22	15	207–255
	080	A or N or NT or QT	90,000	50,000	20	35	163–207
	090	NT or NQT	105,000	60,000	17	40	187–241
	0105*	NQT	120,000	85,000	14	35	217–248
	0120*	NQT	150,000	95,000	9	30	248–311
	0150*	NQT	175,000	125,000	6	22	311–363
	0175*	NQT		145,000		12	363–415
AAR M 201–62	A	Unannealed	60,000	30,000	22	30	—
	A	A or N	60,000	30,000	26	38	—
	B	A or N	70,000	38,000	24	36	—(10)
	C	NT or QT	90,000	60,000	22	45	—
	D	QT	105,000	85,000	17	35	—
	E	QT	120,000	100,000	14	30	—
A.B.S.	1	A or NT	60,000	30,000	24	35	Bend—Degrees 120
	2	A or NT	70,000	36,000	22	30	90
	Hull	A or NT	60,000	30,000	24	35	120

American Specifications for Steel Castings—*continued*

CARBON AND LOW ALLOY CAST STEELS—*continued*

Composition

Specification	C	Mn	P	S	Si	Ni	Cr	Mo	Other Elements
				CHEMICAL COMPOSITION % —MAXIMUM					
SAE Automotive	·12–·22	·50–·90	·05	·06	·60	—	—	—	—
	·25(2)	·75(2)	·05	·06	·80	—	—	—	—
	·30(2)	·70(2)	·05	·06	·80	·50(2)	·25(2)	—	W ·10+Cu ·50(2)
	·40–·50	·50–·90	·05	·06	·80	—	—	—	—
	·40–·50	·50–·90	·05	·06	·80	—	—	—	—
	—	—	·05	·06	—	—	—	—	—
	—	—	·05	·06	—	—	—	—	—
	—	—	·05	·06	—	—	—	—	—
	—	—	·05	·06	—	—	—	—	—
AAR M 201–62	—	·85	·05	Basic ·05	—	—	—	—	—
	—	·85	·05	Acid ·06	—	—	—	—	—
	·35	·85	·05	·06	—	—	—	—	—
	—	—	·05	—	—	—	—	—	—
	—	—	·05	—	—	—	—	—	—
A.B.S.	—	—	—	—	—	—	—	—	—
	—	—	—	—	—	—	—	—	—
	—	—	—	—	—	—	—	—	—

Physical Properties

Specification	Class	Heat Treatment #	Tensile Strength p.s.i.	Yield Point p.s.i.	Elong. in 2" %	Red. of Area %	Other Tests: Bend, Impact Hardness†
Lloyds	—	A	26-35(4)	—	20	—	120
	—	A	28-35(4)	—	20	—	120
	—	A	28-35(4)	—	TS+E = 57	—	180
	—	A	28-35(4)	—	20	30	120
	—	A	28-35(4)	—	20	30	120
	—	A	28-40(4)	—	20	30	120
	—	A	30-45(4)	—	20	30	120

EXPLANATION OF SYMBOLS

( ) Permitted only if agreed upon by purchaser.

\# Heat Treatment : A—Full Annealed ; N—Normalised ; T—Tempered ; Q—Liquid Quenched.

•* Hardenability requirements when specified.

† Hardness tests when specified in contract or order.

[ ] Figures in brackets are expected values only. No test required unless specified.

(2) Alloying elements shall be selected by the manufacturer.
For each reduction of ·01% C below the max. specified, an increase of ·04% Mn above the max. specified will be permitted to a max. of 1·0% Mn (1·40 max. for grade 70–40, A27).

(3) Total max. content of undesirable elements is 1·0%. For each ·10% below the specified max. alloy content of 1·0%, an increase of ·02% in the Cr plus Mo content and ·06% in the Ni and Cu contents above the max. will be permitted.

(4) Tons (2240) per sq. in.

(5) Total max. content of undesirable elements in 1·0%.

American Specifications for Steel Castings—continued

CARBON AND LOW ALLOY CAST STEELS—continued

Composition

Specification	C	Mn	P	S	Si	Ni	Cr	Mo	Other Elements
Lloyds	—	—	·05	·05	—	—	—	—	
	—	—	·05	·05	—	—	—	—	
	—	—	·05	·05	—	—	—	—	
	·25	·5 – ·10	·05	·05	·15–·5	—	—	—	
	·25	1·0–1·5	·05	·05	·15–·5	—	—	—	
	·25	·5–1·0	·05	·05	·15–·5	—	·4 – ·7		
	·25	·5– ·8	·05	·05	·15–·5	—	·8–1·4	·45–·65	

(6) Restrictions on unspecified alloy elements : W ·10% for all grades ; total max. content of unspecified elements 1·0% for all grades except WC4 and WC5 where total max. is ·60%.
(7) Charpy keyhole notch impact test required at temperature specified by customer.
(8) Vanadium 0·01% minimum may be specified or up to 0·05% permitted.
(9) Vanadium 0·05% minimum may be specified.
(10) Hardenability requirement of Rc=40 max. at 10/16″.
(11) For each reduction of ·01% below the specified maximum C content, an increase of ·04% Mn above the specified max. will be permitted up to a max. of 1·1%.
(12) The use of aluminium is prohibited in the making of all steels except grade I.
(13) Proprietary Steel Composition.
(14) Charpy V-Notch at 70° F. Values apply only to sections to 2 inches.
(15) Restriction on unspecified elements Cu 0·50, W 0·10, total 0·60%.
(16) Restriction on unspecified elements Cu 0·50, W 0·10, total 1·00%.

342

HIGH ALLOY CAST STEELS

Physical Properties

	SPECIFICATION & HEAT TREATMENT		MECHANICAL PROPERTIES—MINIMUM				Other Tests: Bend, Impact Hardness (150)
Specification	Class	Heat Treatment #	Tensile Strength p.s.i.	Yield Point p.s.i.	Elong. in 2" %	Red. of Area %	Bend—Degrees
ASTM A-128-60	—	Q	—	—	—	—	—
ASTM A296-63T	CA-15	NT or A	90,000	65,000	18	30	BHN-241 max.
	CA-40	NT or A	100,000	70,000	15	25	—
	CB-30	N or A	65,000	30,000	—	—	BHN-241 max.
	CC-50	N or A	55,000	—	—	—	BHN-241 max.
	CE-30	—*	80,000	40,000	10	—	—
	CF-3	—*	65,000	28,000	35	—	—
	CF-3M	—*	70,000	30,000	30	—	—
	CF-8	—*	65,000	28,000	35	—	—
	CF-8C(3)	—*	70,000	30,000	30	—	—
	CF-8M	—*	70,000	30,000	30	—	—
	CF-16F(2)	—*	70,000	30,000	25	—	—
	CF-20	—*	70,000	30,000	30	—	—
	CG-8M	—*	75,000	35,000	25	—	—
	CG-12	—*	70,000	28,000	35	—	—
	CH-20(2)	—*	70,000(7)	30,000	30	—	—
	CK-20	—*	65,000	28,000	30	—	—
	CN-7M	—*	62,500	25,000	35	—	—

HIGH ALLOY CAST STEELS—continued

Composition

Specification	CHEMICAL COMPOSITION %—MAXIMUM								Other Elements
	C	Mn	P	S	Si	Ni	Cr	Mo	
ASTM A 128-60	1·00-1·40	10·0(1)	·10	·05	—	—	—	—	—
ASTM A 296-63T	·15	1·00	·04	·04	1·50	1·00	11·5-14	·50	
	·20-·40	·40-1·00	·04	·04	1·50	1·00	11·5-14	·50	
	·30	1·00	·04	·04	1·00	2·00	18-21	—	(4)
	·50	1·00	·04	·04	1·00	4·00	26-30	—	
	·30	1·50	·04	·04	2·00	8-11	26-30	—	
	·03	1·50	·04	·04	2·00	8-12	17-21	—	
	·08	1·50	·04	·04	1·50	9-11	17-21	2-3	Cb(5)
	·03	1·50	·04	·04	2·00	8-11	18-21	—	
	·08	1·50	·04	·04	2·00	9-12	18-21	2-3(2)	Se(2)
	·08	1·50	·04(2)	·04(2)	2·00	9-12	18-21	—	
	·16	1·50	·04	·04	2·00	9-12	18-21	—	
	·20	1·50	·04	·04	1·50	8-11	18-21	3-4	
	·08	1·50	·04	·04	2·00	9-13	18-21	—	
	·12	1·50	·04	·04	2·00	10-13	20-23	—	
	·20(3)	1·50	·04	·04	2·00	12-15	22-26	—	
	·20	2·00	·04	·04	2·00	19-22	23-27	—	Cu 3-4
	·07	1·50	·04	·04	1·50	27·5-30·5	19-22	2-3	

Physical Properties

SPECIFICATION & HEAT TREATMENT			MECHANICAL PROPERTIES—MINIMUM				
Specification	Class	Heat Treatment #	Tensile Strength p.s.i.	Yield Point p.s.i.	Elong. in 2" %	Red. of Area %	Other Tests: Bend, Impact Hardness
ASTM A297-63	HC	—	55,000	—	—	—	—
	HD	—	75,000	35,000	8	—	—
	HE	—	85,000	40,000	9	—	—
	HF	—	70,000	35,000	25	—	—
	HH	—	75,000	35,000	10	—	—
	HI	—	70,000	35,000	10	—	—
	HK	—	65,000	35,000	10	—	—
	HL	—	65,000	35,000	8	—	—
	HN	—	63,000	—	4	—	—
	HT	—	65,000	—	4	—	—
	HU	—	65,000	—	—	—	—
	HW	—	60,000	—	—	—	—
	HX	—	60,000	—	—	—	—
ASTM A351–63T	CA15	NT	90,000	65,000	18	30	—
	CF3	N or Q	70,000	30,000	35	—	—
	CF8	N or Q	70,000	30,000	35	—	—
	CF3M	N or Q	70,000	30,000	30	—	—
	CF8M	N or Q	70,000	30,000	30	—	—
	CF8C	N or Q	70,000	30,000	30	—	—
	CH8	N or Q	65,000	28,000	30	—	—
	CH10	N or Q	70,000	30,000	30	—	—
	CH20(8)	N or Q	70,000	30,000	30	—	—
	CT35	N or Q	65,000	28,000	15	—	—
	CF10MC	N or Q	70,000	30,000	20	—	—
	CN7M	N or Q	62,500	25,000	35	—	—
	CK45	N or Q	65,000	35,000	10	—	—

American Specifications for Steel Castings—continued

HIGH ALLOY CAST STEELS—continued

Composition

Specification	C	Mn	P	S	Si	Ni	Cr	Mo	Other Elements
						CHEMICAL COMPOSITION % —MAXIMUM			
ASTM A 297–63	·50	1·00	·04	·04	2·00	4·00	26–30	·50[8]	
	·50	1·50	·04	·04	2·00	4–7	26–30	·50[8]	
	·20–·50	2·00	·04	·04	2·00	8–11	26–30	·50[8]	
	·20–·40	2·00	·04	·04	2·00	8–12	18–23	·50[8]	
	·20–·50	2·00	·04	·04	2·00	11–14	24–28	·50[8]	
	·20–·50	2·00	·04	·04	2·00	14–18	26–30	·50[8]	
	·20–·60	2·00	·04	·04	2·00	18–22	24–28	·50[8]	
	·20–·60	2·00	·04	·04	2·00	18–22	28–32	·50[8]	
	·20–·50	2·00	·04	·04	2·00	23–27	19–23	·50[8]	
	·35–·75	2·00	·04	·04	2·50	33–37	13–17	·50[8]	
	·35–·75	2·00	·04	·04	2·50	37–41	17–21	·50[8]	
	·35–·75	2·00	·04	·04	2·50	58–62	10–14	·50[8]	
	·35–·75	2·00	·04	·04	2·50	64–68	15–19	·50[8]	
ASTM A 351–63T	·15	1·00	·04	·04	1·50	1·00	11·5–14	·50	
	·03	1·50	·04	·04	2·00	8–12	17–21		
	·08	1·50	·04	·04	1·50	8–11	18–21		
	·03	1·50	·04	·04	1·50	9–12	17–21	2–3	
	·08	1·50	·04	·04	2·00	9–12	18–21	2–3	
	·08	1·50	·04	·04	1·50	9–12	18–21		[5]
	·08	1·50	·04	·04	2·00	12–15	22–26		
	·10	1·50	·04	·04	1·50	12–15	22–26		
	·20[9]	2·00	·04	·04	2·50	12–15	22–26		
	·35	1·50	·04	·04	2·50	33–37	13–17	·50	
	·10	2·00	·04	·04	1·50	13–16	15–18	1·75–2·25	[10]
	·07	1·50	·04	·04	1·50	27·5–30·5	19–22	2–3	Cu 3–4
	·20–·45	1·50	·04	·04	1·75	23–27	19–22		

Physical Properties

	SPECIFICATION & HEAT TREATMENT		MECHANICAL PROPERTIES—MINIMUM				
Specification	Class	Heat Treatment #	Tensile Strength p.s.i.	Yield Point p.s.i.	Elong. in 2" %	Red. of Area %	Other Tests: Bend, Impact Hardness
ASTM A447–50	I	—	80,000[11]	—	9	—	M. Perm. 1.70 Rupt. 5000 psi
	II	—	— 80,000[11] 20,000[12]	— — —	— 4 8	— — —	M. Perm. 1.05 Rupt. 8000 psi
ASTM A448–50	—	—	65,000	—	4	—	—
MILITARY MIL–S –16993A	1 2	NT NT	90,000 90,000	65,000[15] 65,000	18 18	30 30	— —
MILITARY MIL–S– 867A	I II III	O O O	70,000 70,000 70,000	28,000 30,000 30,000	35 30 30	— — —	— — —

347

American Specifications for Steel Castings—continued

HIGH ALLOY CAST STEELS

Composition

Specification	CHEMICAL COMPOSITION %—MAXIMUM								
	C	Mn	P	S	Si	Ni	Cr	Mo	Other Elements
ASTM A447-50	.20–.45 .20–.45	2·50 2·50	·05 ·05	·05 ·05	1·75 1·75	10–14[13] 10–14[13]	23–28 23–28	— —	N ·20 N ·20
ASTM A448-50	·35[14]–·75	·50–2·00	·04	·04	·80–2·50	33–37[13]	14–17	—	Ni + Cr 48 min.
MILITARY MIL–S–16993A	·15 ·15	1·00 1·00	·05 ·05	·05 ·05	1·50 ·50	1·00 ·65–1·0	11·5–14·0 11·5–14·0	·50 ·50–·70	
MILITARY MIL–S–867A	·08[14] ·08 ·08	1·50 1·50 1·50	·05 ·05 ·05	·05 ·05 ·05	2·00 2·00 2·00	8–11 9–12 9–12	18–21 18–21 18–21	— — 2–3	Cb + Ta 1·1[17]

• Heat treatment as agreed upon by manufacturer and purchaser.

(1) Mn—Minimum per cent.

(2) For free machining properties the composition of grade CF-16F may contain suitable combinations of Se, P, and Mo (grade CF-16F) or of S and Mo (grade CF-16Fa) as follows :

Selenium, phosphorus, and molybdenum :   Sulfur and molybdenum :

Selenium, %	·20 to 35	Sulfur, %	·20 to ·40
Phosphorus, max., %	·17	Molybdenum, %	·40 to ·80
Molybdenum, max., %	1·50		

(3) For more severe general corrosive conditions, and when so specified, carbon content shall not exceed ·10%. Low carbon grade shall be designated as grade CH-10.

(4) Chemical analysis not normally required for P, S, and Mo, but if present in amounts over those stated may be cause for rejection.

(5) Grade CF-8C shall have a columbium content of not less than eight times the carbon content and not more than 1·0 per cent.

(6) For grade CB-30 a copper content of ·90 to 1·20% is optional.

(7) A tensile strength of 65,000 psi min. is permitted when the carbon content is ·06% max., or the silicon content is 1·00% max. or both.

(8) Castings having a specified molybdenum range agreed upon by the manufacturer and the purchaser may also be furnished under these specifications.

(9) By agreement, the carbon content of grade CH20 may be restricted to ·10% maximum. The grade designation shall be CH10.

(10) Grade CF10MC shall have a columbium content of not less than 10 times the carbon content but not over 1·20%.

(11) Properties after aging.

(12) Short-Time High-Temperature Properties at 1600° F. ± 10° F.

(13) Commercial nickel usually carries a small amount of cobalt, and within the usual limits cobalt shall be counted as nickel.

(14) Within carbon range, manufacturer and purchaser shall agree upon nominal carbon content of not less than ·45% nor more than ·65%. Tolerance on such specified carbon content shall be ± 0·10%.

(15) 0·2% offset.

(16) If chromium is over 20% and nickel is over 10%, a maximum carbon content of ·12 will be permitted.

(17) Columbium or columbium plus tantalum shall be not less than 10 times the carbon content and not more than 1·10% (tantalum shall not exceed 4 times the sum of the columbium and tantalum content) or titanium content shall be not less than 6 times the carbon content and not more than ·75%.

SUMMARY OF COMPOSITIONS, TENSILE

Specification	Grade	Type of steel	Chemical				
			C	Si	Mn	Ni	Cr
B.S. 592	A	Carbon steel	0·25 max.	0·60 max.	1·00 max.	0·40 max.*	0·25 max.*
	B	Carbon steel	0·35 max.	0·60 max.	1·00 max.	—	—
	C	Carbon steel	0·45 max.	0·60 max.	1·00 max.	—	—
B.S. 1398		Carbon-molybdenum steel	0·15/0·25	0·20/0·50	0·50/1·00	0·40 max.*	0·25 max.*
B.S. 1456	A	1½ per cent manganese steel	0·18/0·25	0·50 max.	1·20/1·70	—	—
	B		0·25/0·33	0·50 max.	1·20/1·70	—	—
B.S. 1457		Austenitic manganese steel	1·0/1·35	1·0 max.	11·0 min.	—	—
B.S. 1458		45/55 ton alloy steel	—	—	—	—	—
B.S. 1459		55/65 ton alloy steel	—	—	—	—	—
B.S. 1461		3 per cent Chromium-molybdenum steel	0·25 max.	0·75 max.	0·30/0·70	0·40 max.*	2·50/3·50
B.S. 1462		5 per cent Chromium-molybdenum steel	0·20 max.	0·75 max.	0·40/0·70	0·40 max.*	4·00/6·0
B.S. 1463		9 per cent Chromium-molybdenum steel	0·20 max.	1·00 max.	0·30/0·70	0·40 max.*	8·00/10·0
B.S. 1617	A	High magnetic permeability mild steel	0·15 max.	0·60 max.	0·50 max.	0·40 max.*	0·25 max.
	B		0·25 max.	0·60 max.	0·50 max.	0·40 max.*	0·25 max.
B.S. 1630	A	13 per cent Chromium steel	0·15 max.	1·00 max.	1·00 max.	1·00 max.	11·5/13·5
	B		0·12/0·20	1·25 max.	1·00 max.	1·00 max.	11·5/13·5
	C		0·20/0·30	1·00 max.	1·00 max.	1·00 max.	11·5/13·5

*Residuals

The above extracts from B.S. 3100 : 1957 are reproduced by permission of the British Standards Institution, 2 Park Street, London, W.I, from whom copies of the complete standard may be purchased.

STRENGTHS AND YIELD STRESSES

					Tensile strength (min. unless a range is indic.)		Yield stress min.	
Mo	Cu max.	Nb or Ti	S max.	P max.	tons/ sq. in.	kg/ sq. mm.	tons/ sq. in.	kg/ sq. mm.
·15 max.*	0·30*	—	0·060	0·060	28	44	14	22
—	—	—	0·060	0·060	32	50	16	25
—	—	—	0·060	0·060	35	55	17·5	28
·40/0·70	0·40*	—	0·050	0·050	30	47	16	25
—	—	—	0·050	0·050	35/45	55/71	60% T.S.	60% T.S.
—	—	—	0·050	0·050	45/55	71/87	60% T.S.	60% T.S.
—	—	—	0·060	0·10	—	—	—	—
—	—	—	0·050	0·050	45/55	71/87	32	51
—	—	—	0·050	0·050	55/65	87/102	38	60
·35/0·60	0·40*	—	0·050	0·050	40/50	63/79	24	38
·45/0·65	0·40*	—	0·040	0·040	40	63	27	43
·90/1·20	0·40*	—	0·040	0·040	40	63	27	43
·5 max.*	0·40*	—	0·060	0·060	23 max.	44	—	—
·5 max.*	0·40*	—	0·060	0·060	26/32	41/50	13	20
·5 max.*	0·40*	—	0·050	0·050	35	55	24	38
5 max.*	0·40*	—	0·050	0·050	40	63	29	46
5 max.*	0·40*	—	0·050	0·050	45	71	30	48

SUMMARY OF COMPOSITIONS, TENS

Specifi-cation	Grade	Type of steel	Chem C	Si	Mn	Ni	Cr
B.S. 1631	A B.Nn B.Ti	Austenitic Chromium- nickel steel	0·12 max. 0·12 max. 0·12 max.	2·00 max. 2·00 max. 2·00 max.	2·00 max. 2·00 max. 2·00 max.	7·00 min. 8·5 min. 7·5 min. Ni+Cr : 25·0 min	17·0 m 17·0/2(17·0/2(
B.S. 1632	A B C.Nb C.Ti D	Austenitic Chromium- nickel- molybdenum steel	0·08 max. 0·08 max. 0·12 max. 0·12 max. 0·08 max.	1·50 max. 1·50 max. 1·50 max. 1·50 max. 1·50 max.	2·00 max. 2·00 max. 2·00 max. 2·00 max. 2·00 max.	11·0/14·0 10·0 min. 10·0 min. 10·0 min. 8·00 min.	18·0/2(16·5/18 16·5/18 16·5/18 16·5/18
B.S. 1648	A B C D E F G H K	Heat resisting alloy steel	0·25 max. 1·0 max. 1·0/2·0 0·4 max. 0·5 max. 0·5 max. 0·5 max. 0·5 max. 0·75 max.	2·0 max. 2·0 max. 2·0 max. 2·0 max. 2·5 max. 3·0 max. 3·0 max. 3·0 max. 3·0 max.	1·0 max. 1·0 max. 1·0 max. 2·0 max. 2·0 max. 2·0 max. 2·0 max. 2·0 max. 2·0 max.	— — — 6·0 min. 10·0/14·0 17·0 22·0 23·0 28·0 36·0/46·0 55·0/65·0	12·0/1(25·0/3(25·0/3(17·0/2: 22·0/2: 22·0/2: 15·0/2(15·0/2 10·0/2(
B.S. 1760	A B	Carbon steel for surface hardening	0·40/0·50 0·55/0·65	0·60 max. 0·60 max.	1·00 max. 1·00 max.	0·40 max.* 0·40 max.*	0·25 max. 0·25 ma
B.S. 1956	A B C	1 per cent Chromium steel (abrasion resisting)	0·45/0·55 0·45/0·55 0·55/0·65	0·75 max. 0·75 max. 0·75 max.	0·50/1·00 0·50/1·00 0·50/1·00	— — —	0·80/1 0·80/1 0·80/1

*Residuals.

STRENGTHS AND YIELD STRESSES

Mo	Cu max.	Nb	Ti	S max.	P max.	Tensile strength (min. unless a range is indic.) tons/ sq. in.	kg/sq. mm.	Yield stress min. tons/ sq. in.	kg/sq. mm.
—	—	8 × C/1·10	—	0·045	0·045	30	47	13·5	21
—	—	—	—	0·045	0·045	30	47	13·5	21
—	—	—	4 × C/0·70	0·045	0·045	30	47	13·5	21
3·00/4·00	—	—	—	0·045	0·045	30	47	13·5	21
2·00/3·00	—	—	—	0·045	0·045	30	47	13·5	21
2·25/2·75	—	8 × C/1·10	—	0·045	0·045	30	47	13·5	21
2·25/2·75	—	—	4 × C/0·70	0·045	0·045	30	47	13·5	21
2·00/3·00	—	—	—	0·045	0·045	30	47	13·5	21
—	—	—	—	—	—	—	—	—	—
—	—	—	—	—	—	—	—	—	—
—	—	—	—	...	—	—	—	—	—
—	—	—	—	—	—	—	—	—	—
—	—	—	—	...	—	—	—	—	—
—	—	—	—	—	—	—	—	—	—
—	—	—	—	—	—	—	—	—	—
0·15 max.*	0·40 max.*	—	—	0·060	0·060	40	63	20	31
0·15 max.*	0·40 max.*	—	—	0·060	0·060	45	71	22	35
—	—	—	—	0·060	0·060	45	71	—	—
—	—	—	—	0·060	0·060	—	—	—	—
0·20/0·40	—	—	—	0·060	0·060	—	—	—	—

F —M

APPLICATION OF
INSULATING AND EXOTHERMIC RISERS
TO CASTINGS

Determination of Exothermic Risers and Pads

The complex design of most castings eliminates the possibility of one universal risering formula, and, furthermore, there are variations between foundries involving melting and pouring practice, temperature control, and even the moulding materials used which must be taken into consideration. A method of conventionally risering castings, which take into account the effect of volume/surface area ratio on freezing time, and the different feed demands of various geometric shapes such as cubes, plates, bars, etc., has been developed.

Similar data have been developed for lined risers, and for the practical foundryman much of the mathematical work has been summarised in simple graphical form, thereby permitting the rapid calculation of correctly proportioned risers.

Graphs, such as shown in Figs. 1-7, indicate the minimum diameters of sleeves which are necessary to feed certain shapes of sections. Then the heights of the sleeves are determined by the weight of the section, and the number of risers on relatively uniform sections are determined by the feed distance from each riser. These basic calculations apply equally to risers lined with either exothermic or insulating sleeves. The thickness of exothermic material required for a given diameter of riser is determined by:

1/10 x diameter of riser + 5 mm.

However, from a practical standpoint this formula is never accurately followed, and generally certain standard wall thickness for a range of riser diameter are used. A typical range is given in Table I.

Table I

Internal diameter (in.)	Wall thickness (in.)	Internal diameter (mm)	Wall thickness (mm)
Below 3	$\frac{1}{2}$	Below 75	12
4–6	$\frac{3}{4}$	100–150	20
7–9	1	175–225	25
10–12	$1\frac{1}{4}$	250–305	31
13–15	$1\frac{1}{2}$	330–380	37
16–18	$1\frac{3}{4}$	405–455	43
19–21	2	480–535	50
22–24	$2\frac{1}{4}$	560–610	56
25–27	$2\frac{1}{2}$	635–685	62
28–30	$2\frac{3}{4}$	710–760	68

Riser Volume

The volume of the riser, as in conventional practice, is determined by the shape factor or cooling rate of the casting. Both for economy and most efficient feeding, the height should be not less than one-half the diameter, nor greater than 1½ times the diameter and it is preferable to regard 1:1 as the ratio for which to aim. The steps involved in risering a casting with sleeves can be summarised as:

1. Determine casting or section weight to be fed.

2. Determine geometric type of casting.

3. Determine from feed distance data, end effect, effect of chill use, etc., the numbers and location of the risers (Table 2).

4. Determine feed metal requirement.

5. Determine the diameter of sleeve necessary to give a sound feed, from the shape classification charts (Figs. 1-7). Where there is a choice, use the larger diameter sleeve.

6. Determine the height of the sleeve, knowing the diameter and weight required (Tables 3 and 4).

7. Determine sleeve thickness required from Table 1 (exothermic sleeves only).

Table 2. Feeding Distances of Risers

Section design thickness	Feeding distance, D_r in terms of casting thickness, T		Square root of T
	Horizontal feeding	Vertical feeding	
	Square bars ($W = T$)		
1	$4T$	$4T$	$1 = 1.00$
2	$4T$	$4T$	$2 = 1.41$
4	$3T$	$3.5T$	$3 = 1.73$
8	$2.1T$	$2.5T$	$4 = 2.00$
12	$1.9T$	$2.0T$	$5 = 2.24$
2 to 12	$6\sqrt{T}$	$6\sqrt{T}$	$6 = 2.45$
			$7 = 2.65$
	Semi-plates ($W = 2T$)		$8 = 2.83$
1	$6T$	$6T$	$9 = 3.00$
2	$6T$	$6T$	$10 = 3.16$
4	$4.5T$	$4.8T$	$11 = 3.32$
8	$4.0T$	$4.5T$	$12 = 3.46$
12	$3.3T$	$3.5T$	
2 to 12	$18\sqrt{T}-16.2$	$16\sqrt{T}-14$	
	Plates ($W = 3T$)		
1	$4.5T$	$4.5T$	
2	$4.5T$	$4.5T$	
4	$4.5T$	$4.5T$	
8	$3.5T$	$3.5T$	
12	$2.8T$		
2 to 12	$11.8\sqrt{T}-5.2$		

A Typical Example using British Units

As a practical example, consider a carbon steel casting in the form of a ring with an outside diameter of 84 in., and with a cross-section of $11\frac{1}{2}$ in. wide x 10 in. deep. Originally this casting was fed by four risers each 18 in. in diameter and 18 in. high. Neglecting the running system and the padding involved (in practice about 6 in. thick at the riser base blended down the casting side for about 8 in.), the total weight of the casting and risers was 12,365 lb with a contact area of about 780 in^2.

The calculations involved and the various considerations taken into account when risering this casting with exothermic sleeves were :

1. *Casting weight.* 7180 lb.

2. *Geometric type of casting.* Since the casting is basically a horizontal square bar bent into a curve, risering and feeding range rules applicable to a bar square section may be used. This was considered for calculation purposes as being 10½ in. square section.

3. *Feeding distance.* Since the casting is circular, end effects which normally increase feeding distances are absent. According to Steel Founders' Society of America rule the feeding distance from the outside diameter of the exothermic sleeve is 1·9 times the casting thickness on either side, the use of four risers amply meets the feed range requirements.

4. *Feed metal requirements.* A fairly heavy sectioned casting of an average type requires about 25% head metal. In the present case the total riser weight is therefore 1795 lb. Since four risers are required, each riser must therefore weigh about 450 lb.

5. *Sleeve diameter.* Figure 6 shows that the minimum riser diameter required to feed a 10½ in. square bar is 10½ in. In practice, were this a relatively short bar with substantial end effects giving additional heat extraction, it would be possible to use a 10 in. diameter head; the residual head would probably not be flat, but slightly dished but the casting would be sound. However, in the ring casting example there is no end effect, and because such castings are usually run by centre sprues, the metal has to flow some distance before entering the mould cavity; the metal entering the riser is therefore often much colder than is ideally suitable for establishing favourable thermal gradients.

Therefore, experience has taught that a slight increase in riser diameter is preferable in these cases, in order to create directional solidification from the casting to the riser. An 11 in. diameter head would therefore be selected as the correct size of head to use in the present case. Since many foundries making their own sleeves have a set of core boxes which gives sleeves in 2 in. diameter increments, when they get a 10 in. diameter and over, a 12 in. diameter head would normally be specified for the present example.

6. *Sleeve height.* Table 3 shows that a 12 in. i.d. sleeve must be at least 14 in. high to contain 450 lb of steel. In practice, 15 in. height is specified to allow for short pouring and hot topping.

7. *Sleeve thickness.* Table 1 shows that the sleeve thickness should be 1½ in.

8. *Further practical considerations.* Following the procedure outlined, it was established that a 12 in. diameter, 15 in. riser would feed the casting. However, from Table 3 it can be seen that a 14 in. diameter, 11 in. high riser would also feed the casting, and it was this riser that was used. This riser was chosen since the exothermic material required to make the sleeves was 28 lb less than for the 12 in. diameter 15 in. high sleeves.

The poured weight when risered conventionally was 12,364 lb, and only 8980 lb when risered with exothermic sleeves. Therefore a weight saving of 3384 lb was realised which was then available for pouring other castings; alternatively, with the same melting capacity, four of these castings could be poured in place of three.

VERTICAL CYLINDER
HEAD TO DIAMETER RATIO.

2 HORIZONTAL CYLINDER.

HEAD TO DIAMETER RATIO.

3 VERTICAL PLATE.

HEAD TO THICKNESS RATIO.

4   HORIZONTAL PLATE.

       HEAD TO THICKNESS RATIO.

5   RECTANGULAR BAR

       HEAD TO THICKNESS RATIO.

DIA. OF FEEDEX HEAD *(INCHES)*

THICKNESS OF BAR *(INCHES)*

DIA. OF FEEDEX HEAD *(CMS)*

THICKNESS OF BAR *(CMS)*

6 SQUARE BAR.

HEAD TO THICKNESS RATIO.

DIA. OF FEEDEX HEAD. *(INCHES)*

THICKNESS OF CUBE. *(INCHES)*

DIA. OF FEEDEX HEAD *(CMS)*

THICKNESS OF CUBE. *(CMS)*

7 CUBE.

HEAD TO THICKNESS RATIO.

362

IMPORTANT NOTES FOR CAST IRONS—
BREAKER CORES

The long freezing ranges of grey irons and, in some cases, malleable, nodular, and high duty irons, are further prolonged by the insulating properties of riser sleeves. This can lead to the development of a coarse grain structure immediately under the head.

Coarsening of the grain structure can be avoided by the use of breaker or knock-off cores, thus restricting feeding necks. Breaker cores must have high dry strength in order to withstand ferrostatic pressures, and must have a low rate of gas evolution on heating.

A core sand base incorporated in a riser shape is by far the most convenient and consistent method of employing breaker cores. The sketch shows a typical riser sleeve of this design.

It will be noted that the sleeve material is blended into the breaker core and the corresponding increase in thickness ensures that no premature freezing off at the neck can occur.

The following table gives the dimensions of breaker cores commonly in use:

Table 3

Diameter of Head		Feeding orifice (circular or square)		Thickness of core	
(in.)	(mm.)	(in.)	(mm.)	(in.)	(mm.)
2·5	63	1·0	25	0·3	8
3·0	76	1·2	30	0·3	8
3·5	88	1·4	36	0·4	10
4·0	102	1·6	41	0·4	10
4·5	114	1·8	46	0·5	12
5·0	127	2·0	51	0·5	12
5·5	139	2·2	56	0·6	15
6·0	152	2·4	61	0·6	15
6·5	164	2·6	66	0·7	18
7·0	178	2·8	71	0·7	18
7·5	190	3·0	76	0·8	20
8·0	203	3·2	81	0·8	20
8·5	215	3·4	86	0·9	23
9·0	229	3·6	91	0·9	23
9·5	241	3·8	97	1·0	25
10·0	254	4·0	102	1·0	25

Table 4
Riser Weight per 6 in. Height

I.D.	Volume	Weight steel	Weight iron	Weight brass	Weight aluminium
(in.)	(in^3)	(lb)	(lb)	(lb)	(lb)
1	5	1	1	1	$\frac{1}{2}$
1½	11	3	3	3	1
2	19	5	5	6	2
2½	29	8	8	9	3
3	42	12	11	13	4
3½	58	16	15	18	5
4	75	21	20	23	7
4½	96	27	25	30	9
5	118	33	31	37	11
5½	143	40	37	44	13
6	170	48	44	53	16
6½	199	56	52	62	18
7	231	65	60	72	21
7½	265	75	69	82	24
8	302	86	79	94	28
8½	340	96	88	105	31
9	382	108	99	118	35
9½	425	120	111	131	39
10	471	133	123	146	43
10½	520	147	135	161	48
11	570	161	148	177	52
11½	623	176	162	193	57
12	679	192	177	210	63
13	796	225	206	247	73
14	924	261	240	286	85
15	1060	300	276	329	98
16	1206	341	314	374	111
17	1362	385	354	422	125
18	1527	432	397	473	140
19	1701	481	442	527	156
20	1885	533	490	584	173
22	2281	646	593	707	210
24	2715	768	706	842	250
26	3186	902	828	988	293

Table 5
Riser Weight per 150 mm Height

I.D.	Volume	Weight steel	Weight iron	Weight brass	Weight aluminium
(mm)	(dm³)	(kg)	(kg)	(kg)	(kg)
30	·10	·83	·76	·85	·27
40	·18	1·47	1·35	1·52	·48
50	·29	2·31	2·12	2·39	·75
60	·42	3·33	3·05	3·43	1·08
70	·57	4·53	4·15	4·67	1·47
80	·75	5·92	5·42	6·10	1·93
90	·95	7·49	6·86	7·72	2·44
100	1·17	9·25	8·48	9·54	3·01
110	1·42	11·20	10·26	11·55	3·65
120	1·69	13·33	12·21	13·74	4·34
130	1·99	15·64	14·33	16·12	5·09
140	2·30	18·14	16·62	18·70	5·91
150	2·65	20·83	19·08	21·47	6·78
160	3·01	23·70	21·71	24·43	7·72
170	3·40	26·76	24·51	27·58	8·71
180	3·81	30·00	27·48	30·91	9·77
190	4·25	33·42	30·62	34·44	10·88
200	4·71	37·04	33·93	38·17	12·06
220	5·70	44·82	41·06	46·19	14·60
240	6·78	53·34	48·86	54·97	17·37
260	7·96	62·60	57·34	64·51	20·39
280	9·23	72·60	66·50	74·82	23·64
300	10·60	83·34	76·34	85·89	27·14
320	12·06	94·83	86·86	97·72	30·88
340	13·62	107·06	98·07	110·33	34·87
360	15·27	120·02	109·94	123·68	39·09
380	17·01	133·73	122·50	137·81	43·55
400	18·85	148·17	135·73	152·70	48·26
450	23·85	187·53	171·78	193·25	61·07
500	29·45	231·52	212·08	238·59	75·40
550	35·64	280·14	256·62	288·70	91·24
600	42·41	333·39	305·40	343·57	108·58
650	49·78	391·27	358·42	403·22	127·43
700	57·73	453·78	415·68	467·64	147·79

The simple mechanism of padding is illustrated in Fig. 8.

In an unheated (unpadded) flat plate, crystals grow in a reasonably symmetrical manner from the mould face and meet at the geometric centre ; at this stage feeding ceases. If an exothermic pad of suitable thickness is placed on one side it will inhibit heat extraction from that side, and the dendrites then have to grow from the unheated surface right the way through the thickness of the plate before feeding ceases and not through half of the thickness as previously. This means in effect that the thickness of the plate has been doubled, and so from Chvorinov's V^2/A^2 rule, its solidification time will be about four times as long.

It also illustrates that there is a practical limit to the extension of solidification time with padding on one side of the section only ; still freezing time depends on the section thickness rather than the exothermic pad thickness provided that the pad is a certain minimum necessary thickness. However, it is possible to increase solidification time more than the four times stated by placing exothermic pads on both sides of the section.

This practical approach is well stated by Wlodawer if, as in Fig. 9, "the junction is led towards the riser by means of correctly dimensioned metal padding, then, because of the thermal similarity of steel and exothermic compounds a wedge of the same thickness can also be made of exothermic material and applied to the casting. However, as soon as the dendrites growing from the unheated face reach the heated face, solidification is completed. It would thus be pointless to make the wedge of exothermic compounds thicker than the casting section, the limit of application of this method has been reached."

Padding Thickness

Although calculations are possible, the severely practical approach has been quite fruitful, and, while some authorities have used pads as thick as the metal section, it has been demonstrated that this is usually unnecessary except in thin sections of 30 mm and under. Some 30 mm sections can be padded with 12 mm thick material adjacent to the riser; and so sections of between 30 mm and 60 mm can be padded with exothermic from 12 to 30 mm thick, dependent upon pad width and the desired length of feeding distance from the riser.

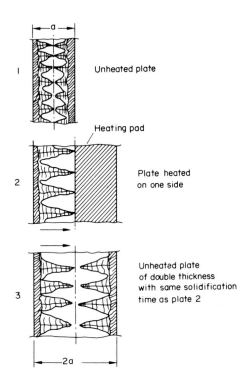

Fig. 8. Effect of exothermic padding on simple solidification pattern (from a paper by Wlodawer, German Foundrymen's Association, Düsseldorf, Autumn, 1961).

As a general guide, for 60—120 mm sections it is usual to consider a pad which is half of the section thickness at the thickest point, tapering to 8 mm thick whatever the length. For thicker sections the pad can be from a quarter to a third the metal section tapering to 8 mm, and so far it has not been found necessary to exceed 130 mm thick pads even on heavy sections.

One important fact is that in conventional practice the weight of metal padding has to be considered as part of the casting, and the riser has to be of such a diameter and weight so as to feed it soundly. The use of exothermic heater pads does not add to the weight of the casting section at all. However, it must be remembered that the thickness of an exothermically padded section is in effect greater than the section, and so allowance must be made in riser diameter.

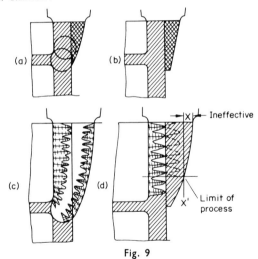

Fig. 9

a. Correctly dimensioned steel padding gives direction of solidification from section junction back to riser (inscribed circles show taper).

b. Location of exothermic pad replacing all steel padding (lower end is thicker because of moulding considerations).

c. Dendrites growing from wall of casting section and from wall of metal pad.

d. Due to rate of growth from unpadded wall, pad any thicker than to line x–x¹ is ineffective.

PADDING FEEDEX

Two types of FEEDEX (61 and 62) are available which can be used in contact with the surface of a steel, SG iron or malleable iron casting, to act as an exothermic pad in place of metal padding.

This padding FEEDEX imparts a consistently good surface finish to the casting, and its exothermic and insulating properties provide effective directional solidification.

FEEDEX 61, mixed with water, and FEEDEX 62, silicate bonded, have excellent green and dry strengths and, being readily mouldable, enable any desired shape to be produced easily.

All types of padding FEEDEX are stable at elevated temperatures and therefore do not deform or allow metal penetration even under high ferrostatic pressures.

Advantages of using Padding Feedex

The main saving from the use of padding FEEDEX is the reduction in fettling costs that is possible compared with the costs incurred in removing conventional metal pads. Figures 1 and 2 illustrate the point; the two malleable iron castings shown are padded with metal and with FEEDEX respectively.

In addition to increasing metal yield, FEEDEX pads provide a consistent, accurately controlled method of feeding. Better profiled castings are produced and it is possible to employ FEEDEX pads in locations where, normally, metal pads would not be considered, due to their inaccessibility.

Incidentally the use of chills and denseners can often be eliminated.

Fig. 1. *Conventional metal padding.*

Fig. 2. *Feedex padding.*

MOULDING AND CORE SAND MIXTURES FOR STEEL CASTINGS

Mixtures developed by the U.S.A. Navy Yard Foundry. The basis is silica sand of 50 to 60 mesh grading.

Green Sand Facing for Steel Castings (proportions by weight)

95% Silica sand.
4% Western bentonite.
1% Cereal type binder.
3% to 4% Water.

The sand and binders are milled dry for five minutes. Then the water is added and milling continued for an additional ten minutes. Old silica sand may be riddled and used to replace new silica sand in proportion up to 50%.

This facing has a green compression strength of about 5 lbs. per square inch, and is suitable for the majority of small and medium weight castings.

Backing Sand for Steel Castings (proportions by weight).

A suitable backing sand contains :—

95% Floor sand.
5% Bentonite.
2·5% to 4% Water.

Dry Sand Facing for Heavy Section Steel Castings

(proportions by weight).

70% Silica sand.
23% Silica flour.
5·5% Bentonite.
1·5% Molasses.
plus 6% to 7% Water.

Dry Sand Facing for Light Section Steel Castings

(proportions by weight).

75% Silica sand.
20% Silica flour.
4% Bentonite.
1% Molasses.
plus 6% to 7% Water.

Core Sand Mixture for Light Section Steel Castings

(proportions by weight).

90% Silica sand.
6% Silica flour.
2% Core oil.
1% Bentonite.
1% Cereal binder.

Core Sand for Small Cores in Average Metal Section

(proportions by weight).

80% Silica sand.
16% Silica flour.
2·5% Core oil.
1·5% Bentonite.

Core Sand for Heavy Section Steel Castings

(proportions by weight).

43·5% Silican sand.
48·5% Silica flour.
4·0% Bentonite.
4·0% Core oil.

NOTE.—All sand mixtures after milling should be put through some form of aerator or disintegrator such as a Royer, before passing to the moulder. This not only improves the permeability but also increases its homogeneity, especially with respect to uniform distribution of moisture. It is also advisable that sand mixtures be prepared some hours before using, as this encourages full development of the bonding properties.

RAMMED REFRACTORY LININGS

ACID STEEL PROCESS

Rammed Lining for Side Blown Convertors

Silica ganister graded minus 15 mm mesh	50%
Silica ganister graded minus 5 mm mesh	30%
Silica flour graded 140 mesh or finer	14%
Western bentonite	6%
Water—by weight	7%

The sieve grading of the dry mixture should be :—

Retained on 28 mesh sieve	...	...	...	...	53–55%
Through 28 and on 65 mesh sieve	...	...	...	12–14½%
Through 65 mesh sieve	...	...	...	...	33–36½%

Mill the ingredients with the water for eight minutes.

After placing in position in the converter, dry slowly at 315° C. up to 36 hours. Heat slowly from 315° C. to 675° C. in from three to four hours. Pre-heat from 675° C. to 1,095° C. before charging in the molten iron.

Rammed Lining for Bottom Pour Steel Ladle

30% Rock ganister.
30% Fireclay.
30% Coarse silica sand.

This mixture is rammed with pneumatic rammers around a metal covered wooden former, which is afterwards withdrawn. It is important that the ganister mixture is added round the former in small quantities and rammed down hard before adding more. This is to ensure the maximum density of the rammed up material.

The bottom is rammed in next, tying in carefully with the side wall material. A clear spot is left in the nozzle area, the nozzle inserted, and the ganister rammed hard around it. Then the bottom is shaped and smoothed over into the nozzle well. Brush over the completed ladle lining with fireclay slurry, set aside, and allow to air dry for at least 48 hours. After this the ladle is dried on low fire at about 540° C. for eight hours. It is subsequently heated to a cherry red 875° C. for at least 15 hours before use.

A high quality ganister rock has the following analysis:

Silica ...	...	...	...	98·82%
Ferrous Oxide	...	...	0·09%
Alumina	...	...	...	0·73%
Magnesia	...	...	...	0·01%
Titania	...	...	...	0·02%
Total alkalies	...	...	0·03%
Ignition loss	...	...	0·30%

A LIST

OF THE

PRINCIPAL

FOSECO PRODUCTS

This list gives a very brief description of, and classifies into groups the principal FOSECO* products according to the metal or alloy for which they are intended, or according to their use. Most products are available in a range of grades to suit varying requirements and conditions, and full details of these grades, and detailed instructions for their use, are given in individual leaflets which are available free, on request.

Brochures, containing the leaflets of each of the above groups, are also available. There is one brochure for each group, and each is distinctively coloured for easy identification.

Your attention is particularly directed to the end of this section where details of FOSECO literature and of the FOSECO Technical Service are given.

*Registered trade name.

SECTION A

PRODUCTS FOR COPPER AND NICKEL
AND THEIR ALLOYS

ALBRAL FLUX

A flux for removing oxides and cleansing copper alloy melts which contain aluminium or silicon. Substantially reduces dross formation and improves fluidity of the metal.

BENTOKOL* Additive

Sand additives for synthetic, semi-synthetic, and some green sand systems. Often eliminates separate additions of coal dust or clay with improved sand properties, better moulding characteristics, and smoother casting finish.

CHROMBAL Flux

Covering fluxes for copper alloys which contain chromium. Reduces the formation of chromium oxide and gas pick-up by the metal during melting.

CUPREX Flux

A range of covering and degassing fluxes for copper and nickel alloys. Provides partial or fully oxidising conditions necessary for minimising hydrogen pick-up during melting.

CUPRIT Flux

Covering and protecting fluxes for application to copper and copper-zinc alloys. Substantially reduces loss of zinc by fuming during melting.

DENSITE Alloying Addition

Alloying additions of nickel and nickel-tin for non-ferrous melts. DENSITE is a convenient means of adding nickel to gunmetal alloys for improving mechanical properties; and to leaded bronze alloys for refining the distribution of lead in the microstructure.

DEOXIDISING TUBES

A wide range of deoxidants for deoxidising and/or increasing the fluidity of copper and nickel alloys. Contains accurate and controlled amounts of deoxidant, thereby making it easy to calculate the amount of deoxidant required for a given melt.

ELIMINAL Flux

Fluxes for the efficient elimination of small quantities of aluminium or silicon from copper and nickel alloys. Removal of these elements automatically increases the fluidity of the metal and improves casting properties generally.

FEEDEX* Anti-piping Compound

Mouldable exothermic anti-piping compounds for the preparation of feeding sleeves to line risers. The exothermic reaction of FEEDEX sleeves vastly increases the feeding efficiency of risers and feeder head size can be substantially reduced.

FEEDOL Compound

Exothermic hot-topping compounds for application to the surface of molten metal in a riser. Prevents heat loss at the surface of the riser and, therefore, improves feeding efficiency. Used in conjunction with FEEDEX exothermic sleeves results in minimum heat loss from risers during solidification of the casting.

INGOTOL* Dressing

Ready bonded chill and ingot dressings for the casting of copper and nickel alloys. Reduces surface defects to a minimum, makes the removal of castings from the moulds easy and facilitates cleaning and dressing of moulds.

KALBORD* Insulating Board

Flexible insulating board for wrapping around a former to line cavities for very large heads.

KALMIN* Sleeves and Shapes

Highly insulating inert riser sleeves and shapes for lining feeder head cavities.

KALMINEX* Shapes

Preformed riser shapes made in a self-heating refractory/ insulating material. Ideal for risers above 12 cm diameter.

LOGAS 50 Degaser

Degassing briquette, which when plunged into copper and nickel alloy melts, removes dissolved hydrogen. Results in significant improvement in mechanical properties.

PLUMBRAL Flux

Covering and cleansing fluxes for copper alloys containing a high lead content. Prevents gas pick-up during melting, reduces the formation of oxide and helps to promote a uniform distribution of lead in the microstructure.

RECUPEX Flux

Range of covering and refining fluxes for the melting of turnings, swarf and other types of scrap of copper and nickel alloy, in all types of furnace. RECUPEX reduces oxidation losses and therefore results in maximum metal recovery. These fluxes also effectively cleanse a melt.

REGENEX Degasser

Powdered cleansing and degassing fluxes for copper and nickel alloys. An alternative to LOGAS 50, particularly for small melts and for charges which consist of fine scrap which is heavily contaminated with non-metallic material.

SLAX* Slag Coagulant

Slag coagulants and controllers to give positive control of slag in the ladle and prevent inclusions in the casting.

TERRADUST Sand Additive

An excellent brass foundry "caster's flour" which does not vary in quality and provides a smooth high quality finish for non-ferrous castings.

*Registered trade mark.

SECTION B

PRODUCTS FOR ALUMINIUM AND
MAGNESIUM AND THEIR ALLOYS

COVERAL Flux
Covering and protecting fluxes for aluminium alloys with types also suitable for the modification of aluminium—silicon alloys.

DEGASER Hydrogen Remover
Tabletted degassing agents for the removal of hydrogen from aluminium melts. Types available to suit all alloys and some combine the action of degassing and grain refining.

DYCASTAL Additive
A compound which, when added in controlled amounts to bale-out ladles, eliminates cracks and draws from aluminium alloy gravity diecastings. Localised shrinkage porosity can also be substantially reduced.

FEEDEX* Anti-piping Compound
See previous Section A.

FEEDOL Compound
See previous Section A.

INERTEX Powder
A dusting compound for protecting molten magnesium alloy during holding and pouring. Provides a barrier at the melt surface which prevents burning of the molten alloy.

KALMIN* 33 Sleeves and Shapes
Highly efficient insulating sleeves and shapes to provide improved heat retention in risers thus increasing feed potential. Do not contaminate metal or sand.

LOMAG* Magnesium Remover
Tabletted compound which, when plunged into molten aluminium alloy, removes small quantities of magnesium where this element is unwanted. With some types of LOMAG, magnesium removal is also accompanied by efficient degassing of the melt.

MAGREX Flux

Covering and cleansing flux for magnesium alloys. Affords efficient protection during melting and, therefore, reduces oxidation to a minimum. On working MAGREX into the melt before pouring, oxides are dissolved and the melt thoroughly cleaned.

NAMETAL Stick Sodium

Foil wrapped metallic sodium for the modification of aluminium—silicon alloys. Available in the form of cubes or sticks.

NAVAC* Metallic Sodium

Vacuum processed metallic sodium packed in air-tight aluminium containers. Sodium in this form is free from all forms of contamination. NAVAC is supplied in accurately weighed units and the degree of modification for a given melt can be carefully controlled.

NUCLEANT Grain Refiner

A range of tabletted grain refining agents for light alloys, including hyper-eutectic silicon aluminium alloys. Grain refining improves the feeding characteristics of an alloy during solidification; and, in the case of heat treatable alloys, mechanical properties are also improved.,

NUCLEANT 200 Grain Refiner

A tabletted grain refining agent which works efficiently at the low temperatures of magnesium alloy melts. Dispenses with the need for super-heating of the melt to obtain grain refinement.

TILITE* Grain Refiner

Tabletted self sinking grain refiner for wrought and casting alloys melted in large bulk melters.

*Registered trade mark.

SECTION C

PRODUCTS FOR IRON AND STEEL

BENTOKOL* Additive
   A composite sand additive giving balanced consistent sand properties eliminating variation and improving casting finish and foundry cleanliness.

BRIX Flux
   Briquettes for adding to the cupola in order to improve fluxing conditions in general. Protects the metal against oxidation during melting and leads to a more efficient refining of the slag. Resulting from the use of BRIX, tapping temperatures are higher and the chances of "bridging" in the cupola are lessened.

BRIXIL* 88 Flux
   A fluxing briquette having a similar action to BRIX, but which also contains a proportion of silicon. A useful product for combining the action of fluxing and raising the silicon content of an iron to improve machinability.

CUPOLLOY* Briquettes
   Briquetted ferro-alloys for adding to the cupola. The briquettes contain a refractory binder which ensures minimum oxidation losses of the alloying additions; they are, moreover, produced in unit quantities so that it becomes unnecessary to break up and weigh ferro-alloys.

FEEDEX* Anti-piping Compound
   See previous Section A.

FERROGEN Deoxidants
   A flux for application to iron in the ladle. FERROGEN has a scavenging action which degasses and removes oxide and slag inclusions from the melt also available as tablets to prevent oxidation in the mould itself.

FERROTUBE Deoxidants
   Having a similar action to FERROGEN but the flux is contained in tubes which are plunged into the melt. These are very convenient for treating small melts.

FERRUX* Anti-piping Compound

Exothermic compound for application to the surface of feeding heads in iron and steel castings. Prevents loss of heat from the surface of risers and, therefore, improves feeding efficiency. Used in conjunction with riser sleeves, it is possible to effect a considerable reduction in the size of feeding heads, thereby economising in melting and fettling costs.

IMPROVITE* Additive

For the simultaneous addition of nickel and copper in the production of high-duty irons. The same beneficial effects as with a pure nickel addition are obtained, but the shot is more easily dissolved. Pressure tightness, machinability and all-round mechanical properties are improved.

INOCULIN and INOTAB* Inoculants

For the efficient ladle or mould inoculation of cast iron melts from high steel charges. A marked increase in tensile and transverse strength is obtained, together with elimination of undercooled graphite. Chilling tendency and section sensitivity of the metal are reduced.

KALMINEX* Shapes

Preformed riser shapes made in a self heating refractory and highly insulating material. Ideal for riser diameters from 12 cm—40 cm.

KALBORD* Insulating Board

Flexible insulating board for positioning in large feeder heads above 40 cm diameter.

LADELLOY* Alloying Agent

Ladle additions of alloying elements. They introduce a uniform weight of element. Alloying·is confined to quantity of metal required.

NICRAD Alloying Addition

For the simultaneous addition of nickel and chromium to the ladle.

SIMCRONITE Inoculant

A combined graphitising and stabilising inoculant for cast iron.

SODA ASH BRIQUETTES

Pure soda ash in block form for convenient addition to furnace or ladle.

SLAX* Slag Coagulant
 A ladle flux for slag coagulation on grey, white or malleable iron.

TELLURIT Dressing
 Mould dressing containing tellurium for producing a chilled layer on the surface of grey iron castings. Allows selective chilling on the surface of an iron casting and uniformity of depth of chill is ensured.

TELLURIUM TUBES
 Pelleted tellurium metal in copper wrapping. For plunging into molten iron to develop uniform hardness and chill in grey iron castings.

*Registered trade mark.

383

SECTION M

MISCELLANEOUS PRODUCTS AND MOULD DRESSINGS

CARSET* Liquid Hardeners

A range of liquid hardeners for use with certain silicates to give an effective, predictable self-setting silicate process.

CARSIL Binders

The FOSECO range of binders for the silicate bonded process of coremaking and moulding.

CHILCOTE Dressing

Dressings for applying to metal chills which are inserted in sand moulds. CHILCOTE prolongs the life of chills, prevents "blowing" from chill surfaces and improves casting surface adjacent to the chill.

CORFIX Adhesive

Core jointing compounds of the self-drying and oven-drying types. Numerous types of CORFIX are available to suit varying requirements. They are easy to use and a minimum application produces a strong permanent joint.

CORSEAL Compound

Diluted with water to form a paste, or in ready to use air-drying form, CORSEAL is recommended for sealing gaps between assembled cores and for patching damaged parts of a core.

DEXIL* Breakdown Additive

Organic breakdown additives for core sand mixes and particularly for sand bonded by the CO_2 process. Produces excellent "shake-out" properties and has less tendency to contaminate the re-used sand than other breakdown additives.

DYCOTE Dressing

A wide range of dressings for application to gravity or pressure die moulds. The range of DYCOTE includes types suitable for light alloy die founding or for the non-ferrous die foundry.

FIRIT and FRACTON* Dressing

A refractory material which, when mixed with water, provides a suitable protective coating for iron and steel furnace tools, ladles, etc. Increases the life of furnace tools and prevents contamination of melts by iron.

FLAMCO Dressing
Powdered mould and core dressings which are ready to use on dilution with isopropyl alcohol or any other suitable alcohol. Particularly suitable for coating CO_2 bonded moulds and cores, FLAMCO is simple to use and easy to prepare.

HARDCOTE Dressings
Dressings for producing a very hard surface to areas of a sand mould. Prevents the eroding action of molten metal during pouring and, therefore, reduces sand inclusions to a minimum. May be used on green sand, dry sand or CO_2 bonded moulds and cores.

HERCULITE PLASTER MC
Casting plaster specially prepared for the production of aluminium matchplates by either gravity or a pressurised casting process.

IPS
Isopropyl alcohol produced by the Shell Chemical Company (not available for export). Recommended for the preparation of FLAMCO and MOLDCOTE products.

INGOTOL* Dressing
Dressings for applying to ingot moulds and notch-bar moulds for steel and non-ferrous casting alloys. Improves the surface finish of the castings and a regular application prolongs the life of moulds.

KALBORD* 20 Insulating Slabs
Highly insulating slabs in segmented form that can be used as "wrap-around" insulation for very large (50 mm and upwards) heads in iron, steel and copper base alloys.

LUBIX Lubricants
Lubricants for the semi-continuous casting of aluminium and its alloys and for plungers of pressure diecasting machines.

MIXAD* Additive
Additives for all clay bonded sands to improve flowability, shatter index, pattern impression and develop full potential of the clay bond: also to prevent casting defects due to sand expansion e.g. buckles, scabs, etc.

MOLDCOTE and CERAMOL* Dressings

Mould and core dressings in paste form or diluted ready to use, containing flammable or self-drying liquid bases. MOLDCOTE makes it unnecessary to torch-dry or stove-dry moulds and cores. In most cases the dressing is ignited after application, whereupon the liquid carrier burns to leave a hard firmly adhering coating to the mould or core.

PATTREX Powder

Pattern stone powders for the preparation of pattern plates, odd sides, Keller models, etc. PATTREX sets with a satin smooth finish and produces a hard and durable surface.

SEPARIT Parting Powder

Superfine, fully water-proofed, non-siliceous parting powders. Ensures a clean lift, even from the most intricate patterns and contains no coarse gritty particles.

SEPAROL Liquid

A liquid parting medium for brushing or spraying on to patterns. Can be used on metal, wood, plastic or plaster patterns. Does not "build up" and promotes a smooth clean casting skin.

TAK* Sealing Compound

A sealing compound for closing gaps between the cope and drag of the mould. Provided in ready to use strips, or in bulk form for preparing strips in the TAK Extruding machine.

TERRACOTE* Dressing

Powdered mould and core dressings for dry application, or, on dilution with water, for wet applications. A wide range is available for all types of mould and core applications and to suit a wide variety of alloys.

TERRAPAINT Dressing

Bonded water-based mould and core dressings for grey iron and steel applications. Available ready to use or as concentrated pastes for water dilution. TERRAPAINT is easy to apply by brush, swab or spray gun.

PLUNGERS

Plunging tools in a variety of lengths, with bell and square-shaped heads, or with hooked and forked ends.

POROTEC* 700 and POROTEC 1000
    Reduced pressure apparatus for determining the relative gas contents of light alloy melts (POROTEC 700) and copper base melts (POROTEC 1000).

RADIANT CORE-VENT Tubing
    A woven synthetic fibre core-vent tubing, available in various diameters.

SPRAY GUNS
    Spray Guns specially designed for spraying MOLDCOTE, FLAMCO, TERRAPAINT, SEPAROL and DYCOTE, etc.

PATENTS
    Many FOSECO products are covered by Patents and/or Patent Applications.

*Registered trade mark.

FOSECO PUBLICATIONS

Several of the leaflets and other publications below are available also in Bulgarian, Czech, French, German, Hungarian, Italian, Japanese, Portuguese, Rumanian, Russian, Spanish and Swedish and may be obtained from the FOSECO representatives in these countries.

LEAFLETS

Leaflets describing the principal FOSECO products and giving full details of the various grades of each, with detailed instructions for their use, are available, free on request. Useful tables, technical data, and in some instances case histories based on practical experience, are also included.

INFORMATION SHEETS

A series of Information Sheets, giving authoritative and up-to-date information and technical data on a wide variety of foundry and allied subjects, are available free, on request.

Information sheet No.	FERROUS
48	Nodular Graphite Cast Iron
49	Malleable Cast Iron
52	Inoculated High Duty Grey Cast Iron
54	High Duty Cast Irons
62	Silicate Bonded Sand
63	The Production of Aluminium Matchplates in Herculite Plaster MC
64	Rimming Steel
66	Cupola Operation Data
67	The Application of Feedex Exothermic Material to Castings

CHARTS

Three Scrap Diagnosis Charts (Ferrous, Non-Ferrous and Light Alloy) listing common casting defects, their appearance, cause, remedy and Foseco products recommended.

A coloured wall chart "The Cupola," shows in very full detail the layout and working principle of the cupola. It is particularly useful for instructing pupils, trainees and foundry personnel generally.

Data charts are available for Iron, Copper Base and Aluminium respectively, giving information on physical properties, melting recommendations and other details for the principal British casting alloys and their equivalents. A further chart describes mould coatings for all types of castings.

FILMS

16 mm films and slides illustrating the correct use and application of various Foseco products are available on free loan to technical societies, branches of the Institute of British Foundrymen, foundries, etc.

INDEX